Ecological Studies, Vol. 222

Analysis and Synthesis

Edited by

M.M. Caldwell, Logan, USA
S.Díaz, Córdoba, Argentina
G.Heldmaier, Marburg, Germany
R.B. Jackson, Durham, USA
O.L. Lange, Würzburg, Germany
D.F. Levia, Newark, USA
H.A. Mooney, Stanford, USA
E.-D. Schulze, Jena, Germany
U. Sommer, Kiel, Germany

Ecological Studies is Springer's premier book series treating all aspects of ecology. These volumes, either authored or edited collections, appear several times each year. They are intended to analyze and synthesize our understanding of natural and managed ecosystems and their constituent organisms and resources at different scales from the biosphere to communities, populations, individual organisms and molecular interactions. Many volumes constitute case studies illustrating and synthesizing ecological principles for an intended audience of scientists, students, environmental managers and policy experts. Recent volumes address biodiversity, global change, landscape ecology, air pollution, ecosystem analysis, microbial ecology, ecophysiology and molecular ecology.

More information about this series at http://www.springer.com/series/86

Daniel G. Gavin • Linda B. Brubaker

Late Pleistocene and Holocene Environmental Change on the Olympic Peninsula, Washington

 Springer

Daniel G. Gavin
Department of Geography
University of Oregon
Eugene, Oregon
USA

Linda B. Brubaker
School of Environmental and Forest Sciences
University of Washington
Seattle, Washington
USA

ISSN 0070-8356   ISSN 2196-971X (electronic)
ISBN 978-3-319-35257-2   ISBN 978-3-319-11014-1 (eBook)
DOI 10.1007/978-3-319-11014-1
Springer Cham Heidelberg New York Dordrecht London

Springer is part of Springer Science+Business Media (www.springer.com)

Preface

Understanding present-day ecosystems must include the concept of time. The fullest understanding of this world, its origin, current status, trends and future will come from the seamless integration of neobiology with paleobiology, and a knowledge of the environmental dynamics that drive the process of change.
– Alan Graham, *A Natural History of the New World* (2011)

The landscape of the Olympic Peninsula in northwest Washington is impressive from every angle. A distinct geographic area bordered by the Pacific Ocean, the Strait of Juan de Fuca, and the Puget Lowlands, the peninsula marks the southern end of a cool-temperate rainforest ecosystem that extends along the archipelago of coastal islands of British Columbia and southeast Alaska. It also contains high-relief mountains, a gradient of rainfall unmatched anywhere in the coterminous USA, forests capable of producing some of the largest and oldest trees in the world, and a distinct biota marked by endemic species. It is this complex geography that has resulted in a unique accumulation and exclusion of biotic diversity on display across the peninsula.

The complex arrangement of ecosystems within the Olympic Peninsula is normally described as a product of the regional climate and geology. However, at a finer resolution, say of a tributary of a major river, one may see that the exact pattern of forest cover is the result of a history of fires in certain years and windthrow in other years (Peterson et al. 1997a). Similarly, if one focuses on a single species, the pattern of its presence and absence over space may be as much a result of environmental gradients as of its recent history of response to disturbances and interactions with other species. Because of the long timescales over which these processes occur on the Olympic Peninsula, making sense of the biotic patterns in this area, or in any geographic area dominated by long-lived species, requires a historical perspective. What events, especially those related to past climate changes and disturbances, have led to the occurrence of the present set of species and its current distribution? Obtaining answers to this question has never been more important than it is today. The rapid climate changes currently underway are unlike any during the last 11,000 years (Marcott et al. 2013), and the potential impact of these climate changes on the biota is difficult to gauge (McMahon et al. 2011).

Changes in climate and vegetation over millennial time scales are fundamental for understanding the assembly of natural communities. Paleoecology, the study of past environments using fossil evidence, not only reconstructs past ecosystems (e.g., species composition, habitat types, etc.) but also addresses ecological questions that cannot be unraveled over timescales of the human lifespan (Schoonmaker and Foster 1991). Paleoecologists tackle several challenges in understanding the biotic response to climate and other agents of ecosystem change that are particularly relevant to the Olympic Peninsula. First, changing disturbance regimes mediates species responses to climate change. Paleoecological studies in the western US have been addressing this issue by using detailed charcoal stratigraphy of fire events in conjunction with the pollen record (Whitlock et al. 2008). On the Olympic Peninsula, many patterns in vegetation can be traced to a history of fire, suggesting an important interaction of fire and climate. Second, current biodiversity hotspots suggest the long-term maintenance of species diversity in these areas. Mountain regions may maintain biodiversity by providing "safe sites" of favorable conditions (refugia) where components of diversity can retreat during unfavorable periods and subsequently expand (Keppel et al. 2012). The modern biogeography of the Olympic Peninsula suggests these mountains were an important refugium south of the ice sheet during the glaciation. Third, the paleoecological approach is especially suited for long-lived organisms. For example, tree species that typically reach reproductive sizes only after 50 years and remain fertile for 300 years, such as many on the Olympic Peninsula, will have experienced only 30–200 generations since colonizing a location after Holocene warming about 11,000 years ago. By summarizing community change through multiple generations and natural disturbance events, paleoecological studies can examine the resilience of ecosystems to disturbances in the past, showing how many ecosystems recover quickly while others may not (Willis et al. 2010).

Interpreting ecological communities and ecosystems from fossil records presents a set of methodological challenges that have been recently addressed with considerable success. Overall, retrospective studies are very labor intensive and knowledge is accumulated in steps, with detailed studies of information-rich sites providing the largest gains. The Olympic Peninsula has no shortage of important paleoecological studies (especially paleovegetation from pollen records) and is nested in a region that has several robust paleoclimate reconstructions.

Aims

This book brings together decades of research on the modern natural environment and paleoenvironmental change since the Late Pleistocene of the Olympic Peninsula. Part 1 presents the modern environment and biogeography of the Olympic Peninsula. Rather than simply reviewing previous studies, we take advantage of a multitude of data sources to portray the gradients of climate, vegetation, and disturbance regimes on the peninsula. Inspired by atlases produced at the University of Oregon (Loy et al. 2001; Marcus et al. 2012), we extensively display primary data in maps and graphics whenever it effectively supplements a point in the text. In Chapter 1, the current climate gradients on the Peninsula and recent trends in climate over the past 100 years demonstrate how climate change is beginning to accelerate on the peninsula. The vegetation of the peninsula is described with respect to five major forest zones and their association with climatic gradients. Lastly, the natural disturbance regime, which is strongly influenced by climate, is reviewed, including fire, wind, insects, and geomorphic disturbances. Chapter 2 presents the geological history and the historical development of biodiversity on the Olympic Peninsula. The long-term geologic history, beginning in the middle Cenozoic, sets the stage for the origin of the Olympic Mountains. What climate changes accompanied the development of the Olympic Mountains during the past 30 million years? How did this region become dominated by some of the largest conifer species in the world? Patterns of endemism and species disjunction are also presented.

In Part 2, we provide temporal depth to the patterns described in the first section. We review the original paleoecological studies from the Pacific Northwest and then turn our focus onto our paleoecological records of changing forest composition and fire over the last 14,500 years. Motivated by pioneering work in addressing ecological questions from the paleovegetation record (Delcourt and Delcourt 1991), we tackle a broad array of ecological questions of the mechanisms underlying the changes in forest communities over space and time. The last 14,500 years begin with the first major warming at the end of the Last Glacial Maximum and the retreat of the continental ice sheet. This period is easily studied due to the abundance of geologic records from lake sediment, ocean sediment, and soil. Over this interval there have been abrupt climate changes, rapid changes in forest composition, and large shifts in fire regimes. Several tree species made their first appearance several millennia after the ice sheet retreated, while others may have been present in small refugia close to the ice-sheet margin, expanding about 13,000 years ago. Tree species that are associated with specific elevations today have shifted their distributions up or down due to changing climate, but sometimes in unexpected ways. The frequency of fire, currently a minor component of the disturbance regime on the Olympic Peninsula, was at times much greater than it is today. Chapter 3 presents the history of postglacial paleoclimate change on the Olympic Peninsula. Various records of past climate are synthesized, including the record of glaciation from the Last Glacial Maximum (21,000 years ago) to the present. Chapter 4 presents the postglacial vegetation history of the biota as determined from a set of pollen records, with five recently developed records receive particular focus. Various means of data synthesis (for example, mapping paleoecological data and modern pollen analog methods) demonstrate changing vegetation patterns. This part of the analysis includes a review of over 30 paleoecological records from western Washington and southwest British Columbia. Chapter 5 presents a brief review of important archeological sites and places their occurrence in time and space in the context of the environmental changes occurring on the peninsula. Lastly, Chapter 6 presents several areas of knowledge gaps and additional needed research.

This book does not explicitly address impacts of ongoing and projected climate changes on the peninsula. There are several assessments related to this topic (Jenkins et al. 2002; Halofsky et al. 2011; Devine et al. 2012). However, the data presented here are relevant to understanding the mechanisms governing ecological change and historical precedents of the modern changing environment on the Olympic Peninsula.

Acknowledgements

Several people have aided in the preparation of this book. Erin Herring conducted an initial literature review and finalized the Wentworth Lake pollen record, Dave Fisher developed excellent R scripts for mapping pollen data, and Ariana White compiled much of the paleoecological data in the report. Dana Lepofsky and Melanie Konradi reviewed portions of the manuscript and Steve Acker reviewed an early draft. We are grateful for advice and information from Steve Acker, Kat Anderson, Allan Ashworth, Patrick Bartlein, Jan Henderson, Robin Lesher, and during initial stages of the project, from Ed Schreiner. Researchers who have shared data include Terri Lacourse, Jan Henderson, Wyatt Oswald, Marlow Pellatt, Susan Prichard, Bruce Bury, Michael Adams, and Chris Ringo. Larry Workman and Gay Hunter generously shared photographs.

We are extremely grateful for the Global Change Research Program to provide funding for paleoecological studies within and around Olympic National Park in 1993–1997. The primary paleoecological data presented in this report represent the efforts of the graduate students in Linda Brubaker's lab. Four Master's students were originally funded through the initial grant in the 1990s (Daniel Gavin, Jason McLachlan, Kyle Young, and Noah Greenwald). In addition to the four M.S. theses, the project resulted in five peer-reviewed publications and one book chapter. Two of these graduate students are now associate professors continuing to work in paleoecology and the other two work in applied and conservation biology.

We are grateful to Dave Conca from Olympic National Park and the Cooperative Ecosystem Studies Unit for providing funding to allow us to focus on completing the analysis of our unpublished data. This project grew into something larger than originally conceived and we are grateful to Dave for allowing us to pursue various directions. Finally, we are grateful to Tim McNulty's research for *Olympic National Park: A Natural History*. McNulty's book provided a springboard for several sections of this book that are afield from our own expertise.

Contents

Part I
The Natural Environment and Biogeography of the Olympic Peninsula, Washington

Chapter 1
The Modern Landscape of the Olympic Peninsula

Abstract The natural environment of the Olympic Peninsula is reviewed with an emphasis on climatic gradients and associated patterns in forest communities and natural disturbance regimes. The regional climate is described with respect to the broadscale circulation features and locally important topographic effects. Analyses of weather station data reveal important trends in temperature, while snow survey and glacier monitoring reveal decades-long decline in snowpack. The summer streamflow of unregulated rivers is shown to have a strong link to variation in snowpack. The seven major vegetation zones of the peninsula are described and controls of species abundance among vegetation zones are discussed with respect to species traits and climatic gradients. The natural disturbance regimes of the peninsula are summarized. Fire history data from tree-ring records and fire regime statistics from historical records reveal the pronounced gradient of fire across the peninsula and the episodic pattern of large fire events. Wind, insect, geomorphic, and intense herbivory are also briefly reviewed.

> Olympic National Park is of remarkable beauty, and is the largest protected area in the temperate region of the world that includes in one complex ecosystems from ocean edge through temperate rainforest, alpine meadows and glaciated mountain peaks. It contains one of the world's largest stands of virgin temperate rainforest, and includes many of the largest coniferous tree species on earth.
> (UNESCO World Heritage Centre, Statement of Significance for Olympic National Park)

1.1 Geography of the Olympic Peninsula and Olympic National Park

The Olympic Mountains encompass roughly 7000 km^2, occupying about 50 % of the Olympic Peninsula in the Pacific Northwest. A large coastal plain occupies the western third of the peninsula, while the mountain massif spans the interior and extends nearly to the north and east coasts. Olympic National Park covers 3696 km^2 of the peninsula, 95 % of which is located on the Olympic Mountains and 5 % located along a coastal strip on the western Olympic Peninsula. Industrial logging has heavily impacted the lowlands circling the park. Fifty-nine percent of the peninsula lies below 350 m in elevation, of which only 6.5 % is protected within the park, whereas 57 % of the area above 350 m elevation is within the park. The park and

© Springer International Publishing Switzerland 2015
D. G. Gavin, L. B. Brubaker, *Late Pleistocene and Holocene Environmental Change on the Olympic Peninsula, Washington,* Ecological Studies 222,
DOI 10.1007/978-3-319-11014-1_1

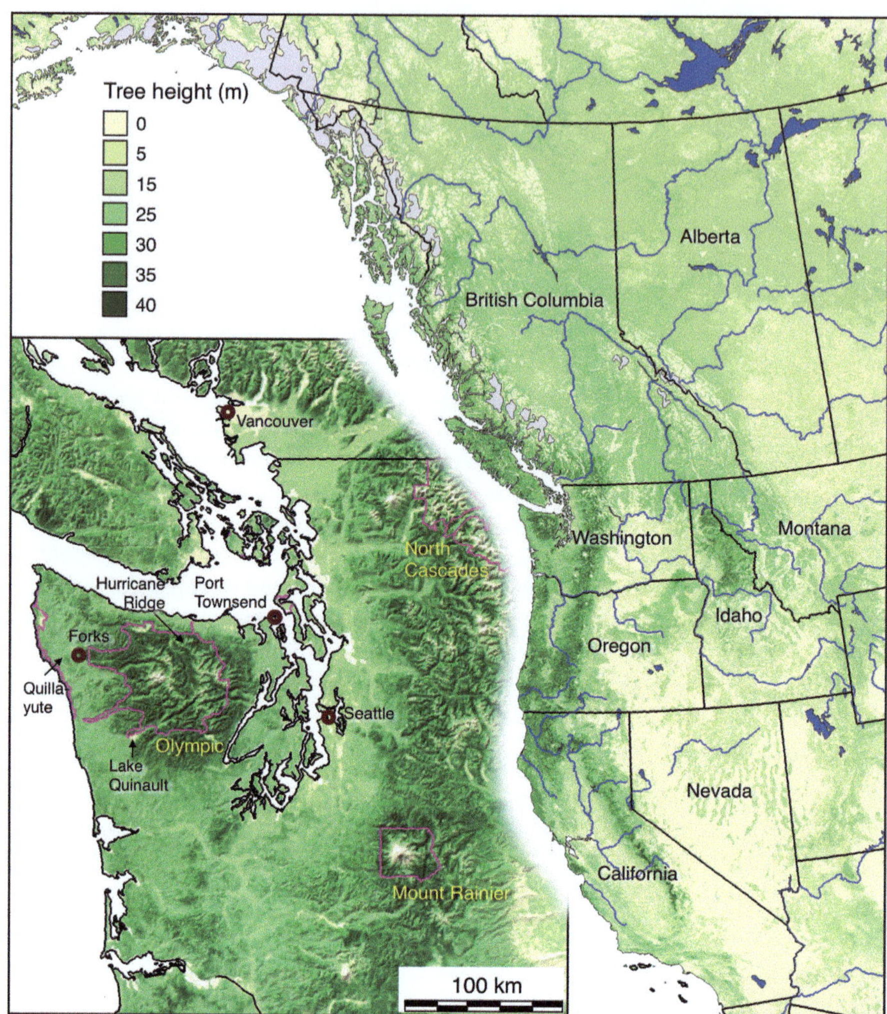

Fig. 1.1 The Pacific Northwest region of North America, centered on Oregon and Washington. Tree height is mapped to show distribution of forests with an emphasis on the tall dense forests of western Oregon, Washington, and British Columbia (Simard et al. 2011). Inset map shows tree height draped over a shaded relief map of western Washington, with national parks (*magenta outlines*) and key locations indicated

surrounding wilderness areas contain one of the largest contiguous intact temperate forests in the world, and the largest contiguous area of unlogged conifer forest in Washington and Oregon (Fig. 1.1).

The Olympic Mountains are a distinct range in the long string of coastal mountains from British Columbia to California. The mountains, which reach an elevation of nearly 2400 m (8000 ft), were formed by the scraping off, or accretion, of ocean-floor sandstone and shale sediments onto the continental margin. Basalt rocks form a distinct horseshoe-shaped band around the north, northeast, and east portion of

Fig. 1.2 Typical forests on a south-facing aspect in the western portion of Olympic National Park, showing an elevation range from 300 m (Queets River at river mile 43) to 1750 m (the summits of Thor and Woden peaks). Forest types include riparian early-successional red alder, avalanche-track slide alder, and old-growth forests of western hemlock, western redcedar, Pacific silver fir, and mountain hemlock. (Photo by Larry Workman)

the mountains. Pleistocene glaciers carved deep U-shaped valleys that extend westward from the interior. These valleys are broadest on the southwest side of the peninsula, whereas valleys on the east and south side are predominantly V-shaped (Fig. 1.2). These features result in a variety of landforms over short distances on the Olympic Peninsula (Fig. 1.3; Montgomery 2002). Modern glaciers, which have been retreating rapidly, are restricted to the highest elevation cirque basins and to the Olympic massif.

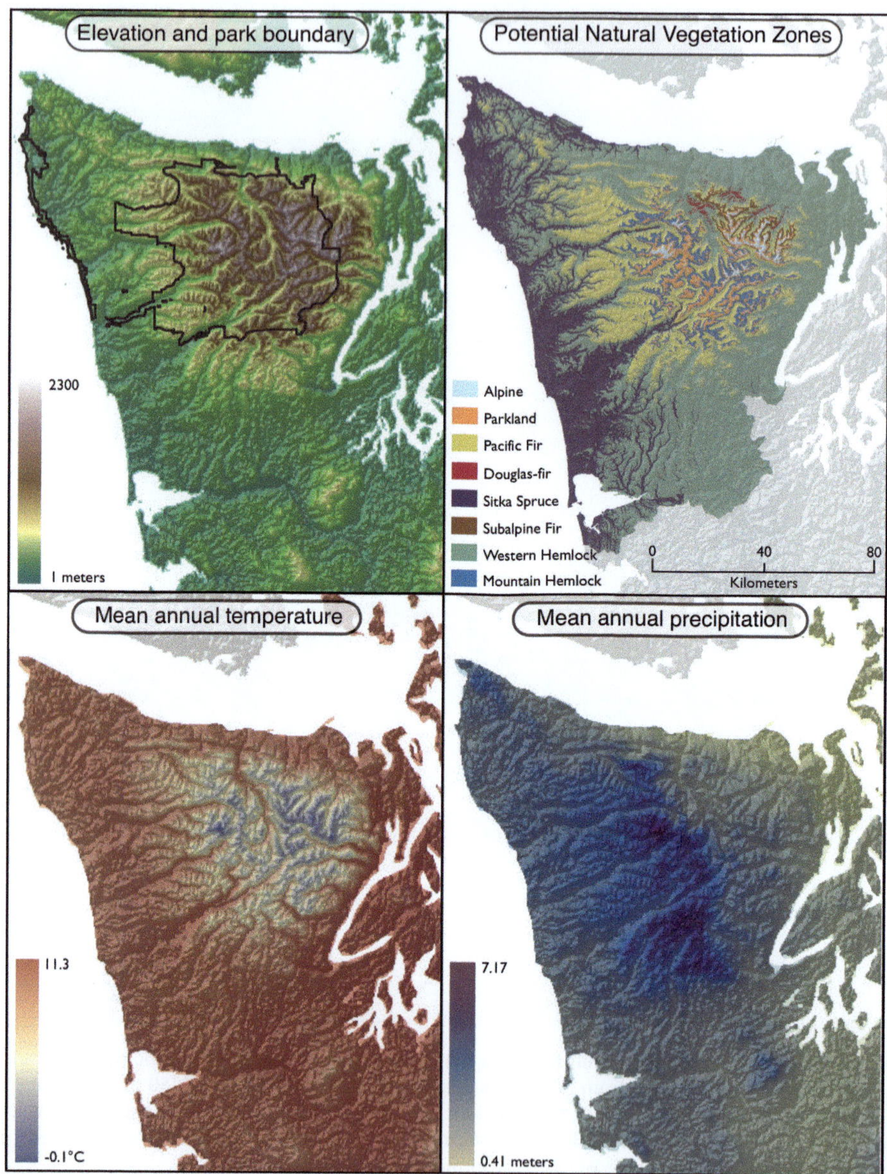

Fig. 1.3 Elevation, vegetation, and climate on the Olympic Peninsula, Washington. Potential natural vegetation zones are from Henderson et al. (2011) and the mean temperature and the mean precipitation are from the PRISM model (Daly et al. 2002). Olympic National Park is shown in the elevation map)

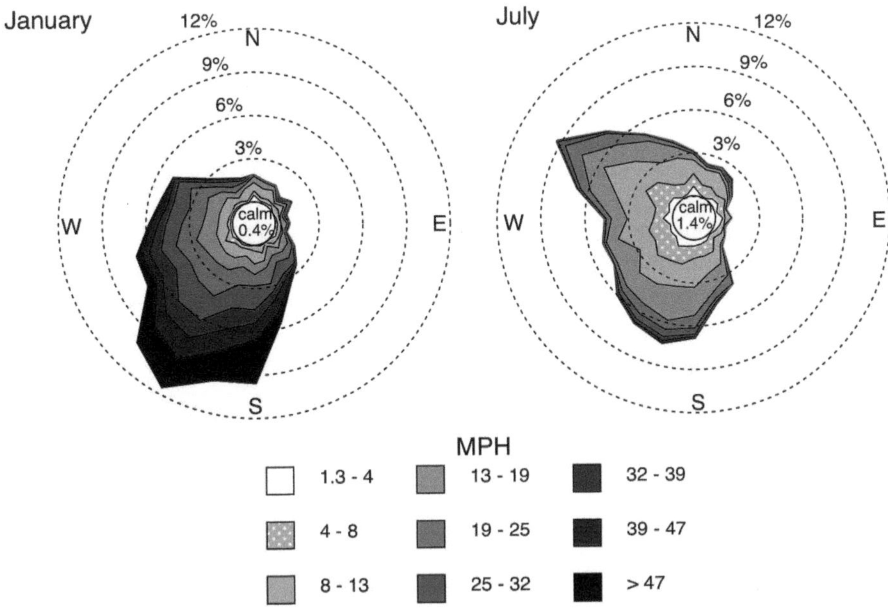

Fig. 1.4 Wind-rose diagrams for 850 mb winds at Quillayute State Airport near Forks, Washington, for the period 1990–2012. The diagrams show the proportion of measurements (twice per day by weather balloon) at a given wind direction and speed. The 850 mb elevation is high enough (generally 5000 ft) such that topography does not influence wind direction and speed. Note the predominately strong SSW winds in January and the gentle NW to SW winds in July. Data were accessed from the NOAA/ESRL radiosonde database (http://www.esrl.noaa.gov/raobs/). *SSW* south-southwest; *NW* northwest; *SW* southwest; *MPH* miles per hour

1.2   Regional Climate on the Olympic Peninsula

The Pacific Northwest climate is characterized by dry summers and wet winters with moderate temperatures year round. This pattern is the result of its midlatitude west-coast location dominated by westerly wind from the Pacific Ocean, where a seasonal pattern of large semipermanent pressure cells exerts the primary control of the patterns of precipitation. During summer, the clockwise flow from a strong Pacific subtropical high-pressure system over the eastern Pacific produces a steady westerly and northwesterly airflow, with descending air resulting in low humidity and clear skies. During the winter, the subtropical high moves southward as the Aleutian low-pressure system intensifies and moves south from the northern Gulf of Alaska. This large low-pressure system picks up moisture over the relatively warm Pacific Ocean and delivers it to the Pacific Northwest. The counterclockwise flow around the Aleutian Low brings air arriving from the south and southwest during the winter, which results in a maritime air mass and moderated winter temperatures (Fig. 1.4). This circulation pattern can, at times, entrain moisture from the subtropics in an "atmospheric river" that can cause intense precipitation sustained for many days.

The Olympic Mountains produce pronounced rainfall gradients at several spatial scales. Moisture-laden air, as it moves inland, is lifted over the mountains and cooled to below the dew-point temperature, producing increased clouds and precipitation at higher elevations. This orographic precipitation occurs on the southwest-facing facets of the major mountain ranges. At a smaller scale, precipitation is greatly affected by individual ridge systems. For example, a recent study has shown that an 800-m-high ridge in the western Olympic Mountains receives 50% more rainfall than the valley bottoms (Anders et al. 2007; Minder et al. 2008). Indeed, the highest precipitation in the coterminous USA, estimated by precipitation models to be more than 5 m, occurs at Mount Olympus (Fig. 1.3). At the largest scale, as soon as air reaches the leeward side of the mountains, it descends and warms, causing rainfall to cease and clouds to dissipate. Because of the consistent southwesterly direction of winter airflow, the southwest portion of the Olympic Mountains receives the majority of precipitation (Fig. 1.3). The rain shadow climate of Sequim, in the northeastern part of the peninsula, receives an average annual rainfall of only 418 mm (1916–1980) although it is located only 55 km from the >5000 mm precipitation at Mount Olympus.

Climate diagrams at five sites summarize diurnal and seasonal patterns of precipitation and temperature across the peninsula from the wet windward side to the dry rain shadow of the northeastern peninsula (Fig. 1.5). The mountains modify temperatures on a daily basis. During the summer, the difference in temperature between the land and the ocean is the greatest, causing a sea breeze from the cool ocean to the warm land, and an upslope breeze into the mountains. Summer maximum temperatures are cooler on the westside (18 °C) than the east side (24 °C) due to the moderating sea breeze on the westside. At night, the breeze reverses in the mountains, bringing a cooler downslope breeze. Summer minimum temperatures are remarkably uniform over the peninsula (ca. 10 °C), likely due to cold air drainages originating in the mountains and reducing air temperature near the ground at night at the low elevations. In the fall, this land breeze can result in valley fog that may take many hours to dissipate each day. The diurnal range of temperature is smaller at high elevation (8 °C in August) than at low elevation (14 °C in August on the eastside) because high elevations receive wind that has not been greatly affected by ground-level warming and cooling. Mean winter temperature is consistently below freezing only in the upper elevations on the eastside of the mountains, though temperatures straddle the freezing mark at the highest elevations on the westside.

1.3 Indicators of Recent Climate Change on the Olympic Peninsula

Over the past 100 years, the greatest climate change on Olympic Peninsula has been an increase in monthly minimum temperatures in the western lowlands (Fig. 1.6). At Forks, both the January and July minimum temperature has increased at a rate of 0.1 °C/decade since 1895, but since 1960 this rate has increased to 0.34 °C/de-

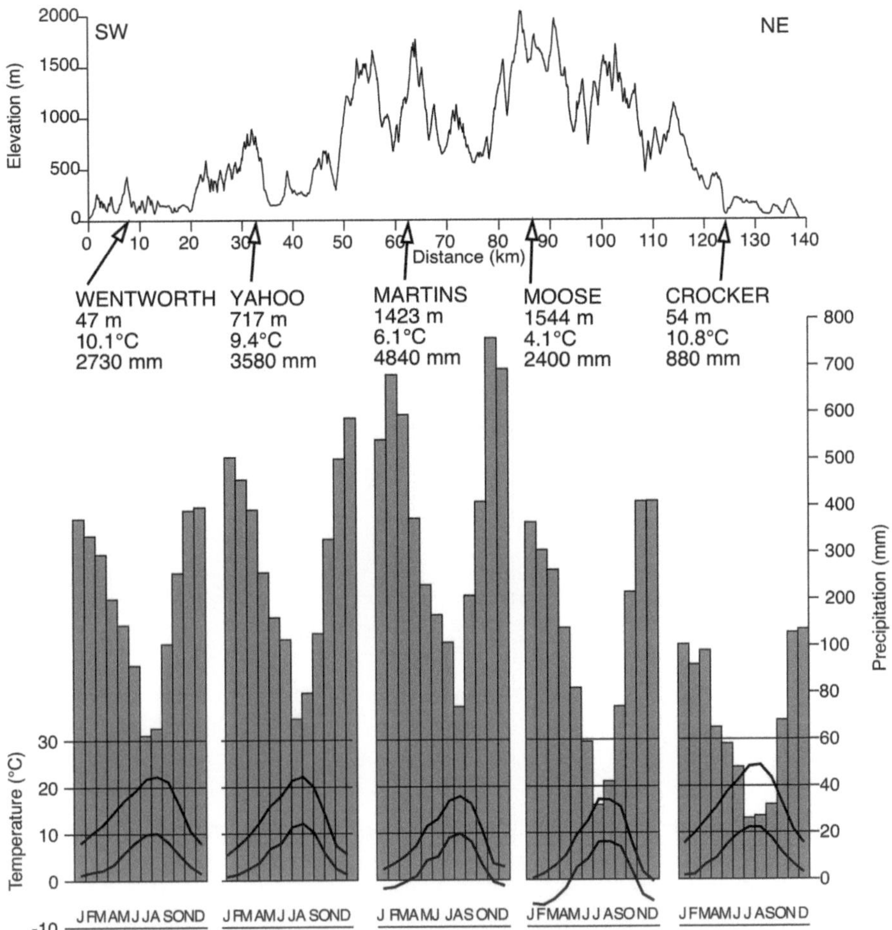

Fig. 1.5 Climate diagrams from five locations straddling the climatic gradient on the Olympic Peninsula. Climate data are 1970–1999 mean monthly precipitation (*grey bars*) and mean monthly maximum and minimum temperature (*black lines*) from PRISM (Daly et al. 2002) and interpolated to local sites using the CLIMATEWNA software (Wang et al. 2012)

cade for January or 0.32 °C/decade for July. At Port Townsend, in the rain shadow, there has been a more muted increase in minimum temperatures since 1960 (0.19 and 0.16 °C/decade for January and July, respectively). At both locations, monthly maximum temperatures do not change more than 0.1 °C/decade. Winter precipitation at Forks increased at a rate of 14.3 mm/decade since 1960, though no change was detected for summer precipitation at Forks and no change in seasonal precipitation was detected at Port Townsend.

Snowpack has decreased dramatically in response to the small increases in winter temperature, because temperatures of the high-snow areas are fairly warm. Nolin and Daly (2006) estimate that 61 % of the snowpack area of the Olympic Mountains

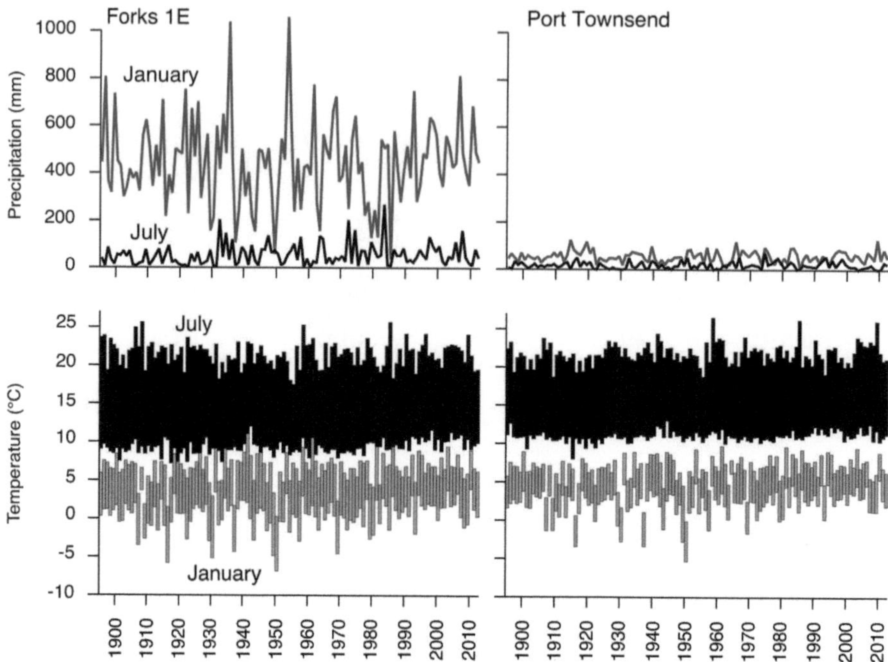

Fig. 1.6 January and July precipitation (*top*) and minimum and maximum temperatures (*bottom*) at Forks and Port Townsend, Washington. Data are from the Historical Climatology Network version 2. (Menne et al. 2009)

is considered "at risk" due to the large quantity of snow occurring at temperatures within 2 °C of the rain–snow boundary. Spring snowpack at Hurricane Ridge in the northern Olympic Mountains declined steadily since the beginning of surveys in 1949 until the late 1990s. It has remained low despite one high-snow year in 1998 (Fig. 1.7). The cause of the decline in snowpack can be attributed to a 2 °C increase in winter minimum temperature on wet days (when rain or snow would occur) between the periods 1948–1976 and 1976–1999 (Conway et al. 1999). This overall trend in snowpack generally follows phases of the Pacific Decadal Oscillation (PDO), a decadal-scale pattern of sea-surface temperature that is strongly correlated to spring temperature and snowfall (Mantua et al. 1997). The interannual variability in snowpack also correlates strongly with the El Niño Southern Oscillation (ENSO); when both ENSO and PDO are in a cold phase, spring snowpack is greater than normal. Regardless of the vagaries of the modes of ENSO and PDO, the data clearly show a link between increasing winter temperature and declining spring snowpack.

The streamflow of the Quinault River, an unregulated river in the southwestern Olympic Peninsula, is an integration of precipitation and early summer snowmelt (Fig. 1.8). The typical annual hydrograph is marked by large winter floods followed by sustained high flow during snowmelt in June and July, with minima occurring in

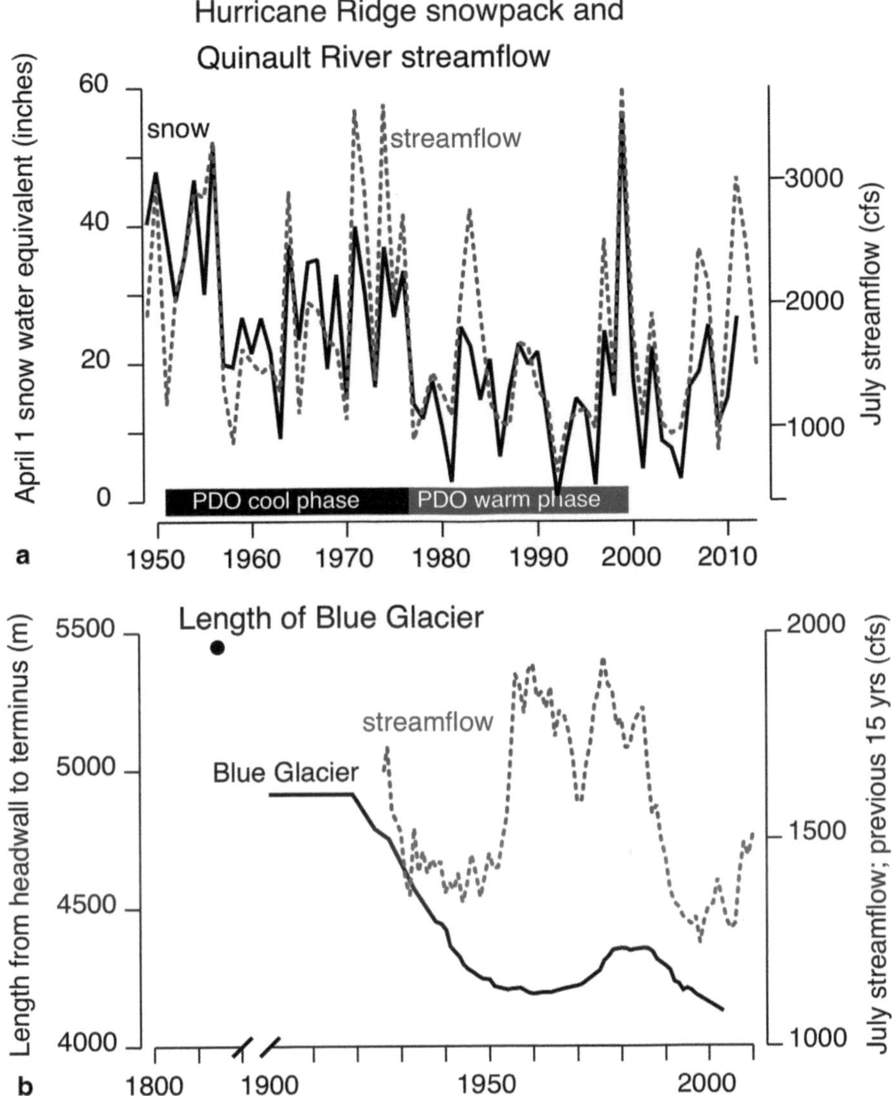

Fig. 1.7 **a** Snow water equivalent on April 1 at Hurricane Ridge snow course from 1949 to 2011 (data from the Natural Resources Conservation Service) and the daily geometric mean for July streamflow of the Quinault River at Quinault Lake. Phases of the Pacific Decadal Oscillation (PDO) are marked by *black and grey lines* (Deser et al. 2004). **b** The length of Blue Glacier from 1815 to 2003 (data from Heusser 1957; Spicer 1989; Conway et al. 1999, and online from the University of Washington) and the mean July streamflow at Quinault Lake as a running geometric mean of the previous 15 years

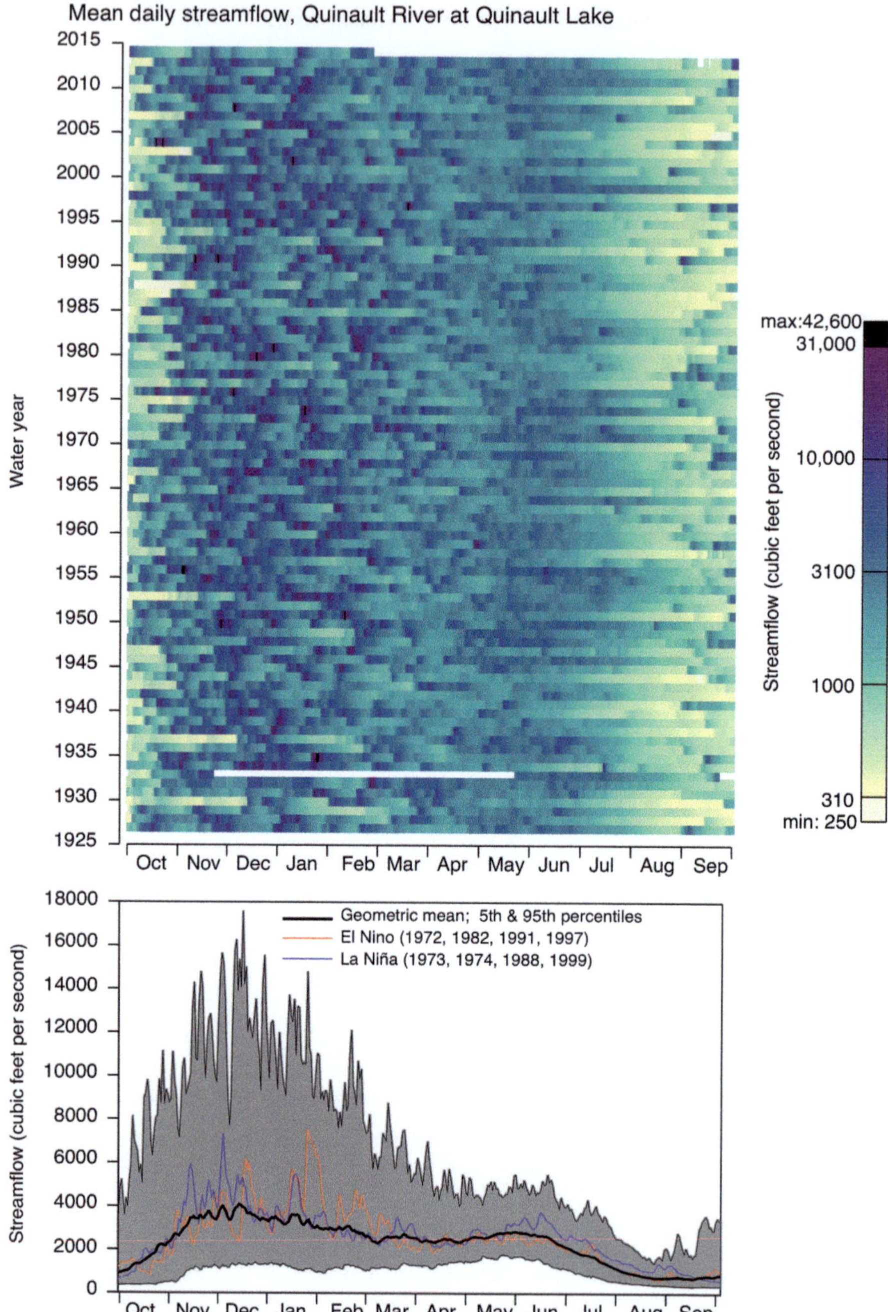

Fig. 1.8 Daily streamflow of the Quinault River at Lake Quinault, 1926–2014. The *lower graph* shows mean daily streamflow (with 5th and 95th percentiles) for all years and for the four largest El Niño and La Niña years

August or September. The four largest El Niño years reveal some high-flow events in midwinter, possibly from rain-on-snow events. In contrast, the four largest La Niña years show a higher sustained summer flow due to a deeper persistent snowpack. These differences in streamflow are consistent with the influence of ENSO on winter climate in the Pacific Northwest.

Summer river flow is strongly dependent on persistent alpine snowpack. April snowpack at Hurricane Ridge in the northeast is strongly correlated with July streamflow on the Quinault River ($r=0.80$; Fig. 1.7). This strong correlation allays some concern regarding a potential bias of the snowpack data. Changing forest cover can affect snowpack accumulation and rates of spring melt, casting doubt on the climatic causes of long-term trends in snowpack. However, the strong streamflow–snowpack correlation (Fig. 1.7) is stationary over time, which would not be the case if the Hurricane Ridge data were not representative of a much larger region.

The history of Blue Glacier on Mount Olympus reveals a longer-term trend of declining snow (Fig. 1.7b). The glacier has retreated almost 1.5 km from its early-nineteenth-century maximum, despite a period of readvance in the 1970s. Similar to causes of declining snowpack, the negative mass balance of recent decades is likely a response to the increased temperature on days with snowfall (Conway et al. 1999). Comparing the Blue Glacier history with July streamflow shows that phases of glacier loss occur during periods of low summer streamflow and phases of glacier stabilization and glacier growth occur during and following periods of high summer streamflow. Further projected warming is expected to accentuate this pattern, reducing spring and early summer streamflow (Halofsky et al. 2011).

1.4 Vegetation Zones of the Olympic Peninsula and Climate Controls on Forest Vegetation

The composition of forests and subalpine meadows on the peninsula has been described in several monographs (Fonda and Bliss 1969; Kuramoto and Bliss 1970; Franklin and Dyrness 1988; Henderson et al. 1989; Schreiner 1994; Buckingham et al. 1995). The descriptions of vegetation "zones" and "series" in this section largely follow Franklin and Dyrness (1988) and Henderson et al. (1989). Recently, Henderson et al. (2011) developed a mapping algorithm to describe the extent of eight "vegetation zones" (Fig. 1.3). These zones broadly follow zones described by previous authors and generally correspond to the dominant late-successional species (Fig. 1.9). Vegetation zones are defined by the dominant vegetation type on a site with a typical soil and drainage, and thus do not capture fine-scale patterns created by such factors as steep slopes or wetlands. However, they do reflect the effect of slope aspect on the temperature regime (Fig. 1.10) and thus forest composition. Henderson et al. (2011) defined vegetation zones using a classification scheme based on a set of nested rules that classify plots using the percent cover of the dominant tree species. The classification scheme was applied to > 1000 vegetation

Fig. 1.9 Four common vegetation zones on the Olympic Peninsula. **a** Mountain hemlock zone forest and parkland, Paradise Valley. Black bear in the foreground. **b** Subalpine fir zone forest and parkland, Obstruction Point. **c** Pacific silver fir zone forest, near Sundown Lake. **d** Western hemlock zone forest, Big Creek. (Photographs (**a**), (**b**), and (**d**) by Larry Workman; photograph (**b**) by Gay Hunter)

plots on the peninsula. These plots were then used in statistical models to develop predictive equations in which environmental variables were used to predict boundary elevations among vegetation zones. These models, based on fog, aspect, cold air drainages, temperature, and precipitation, resulted in an accuracy of ca. 70 % for predicting the correct vegetation zone.

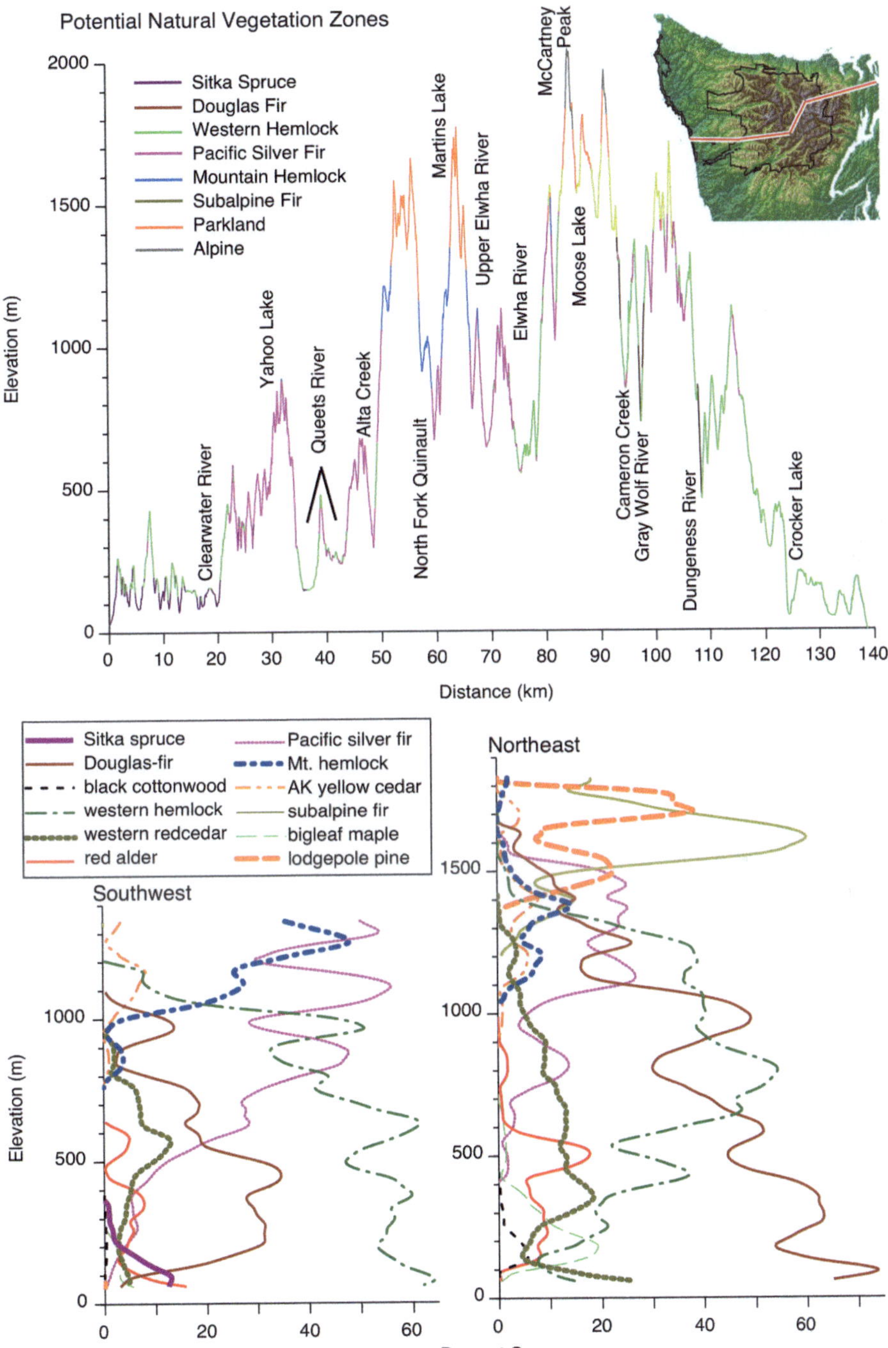

Fig. 1.10 *Top*: An elevation profile across the Olympic Peninsula showing the pattern of potential natural vegetation zones (Henderson et al. 2011). Note how the elevation of zonal boundaries varies with the effects of aspect, cold air drainages in narrow valleys, and location along tran-

The tree species on the Olympic Peninsula have very large geographic ranges, all of which extend into warmer and colder climates that occur on the Olympic Peninsula (Table 1.1). Therefore, the controls on species distributions within the Olympic Peninsula, including the orderly layering of the vegetation zones, must reflect other aspects of the seasonal climate than temperature alone. In an attempt to address the climatic controls of each vegetation zone, we include below a new analysis of climate within each vegetation zone. We focus on the mean temperature and precipitation as well as a simple water balance model, including an index of productivity (actual evapotranspiration) and drought stress (deficit). These latter variables have been shown to be very important controls of tree species distribution (e.g., Gavin and Hu 2006). We also present a short literature review of climatic effects on particular life history stages of the dominant tree species. Relevant studies, such as dendrochronology and seedling recruitment, relate important demographic parameters to climate.

1.4.1   Sitka Spruce Zone

The Sitka spruce zone occupies the low-elevation coastal areas, rarely extending above 250 m elevation, that experience a hypermaritime climate. Fires are very infrequent, with windthrow disturbances more common. Prior to the logging era, forests were dominated by the late-successional species: Sitka spruce, western hemlock, and western redcedar. Common shrubs include salal, salmonberry, red huckleberry, and vine maple. Herbaceous species include oxalis, swordfern, lady-fern, deerfern, and foamflower. Understory regeneration of hemlock, spruce, and redcedar is normally restricted to "nurse logs" that can supply a source of moisture, which is unavailable to the short roots of tree seedlings when in competition with dense moss mats on the forest floor (Harmon and Franklin 1989). River terraces along the Hoh River have been interpreted as a long-term chronosequence (Fonda 1974). Early-successional communities on river floodplains are rapidly colonized by red alder and willows (Van Pelt et al. 2008), but Sitka spruce also establish amidst the hardwoods and eventually replace the hardwoods when they die out after 80 years (Stolnack and Naiman 2010). Red alder is particularly important for building soil nitrogen pools (Luken and Fonda 1983). On upland sites, early-successional communities following wind or the rare fire may be followed by a dense shrub phase, including salmonberry and red elderberry. As Sitka spruce is less shade tolerant than western hemlock, it requires moderate-sized canopy gaps for it to reach the canopy (Taylor 1990). The oldest river terraces are dominated by western hemlock and western redcedar.

There are three distinct variants within the Sitka spruce zone. First, in the major river valleys (Hoh, Quinault, and Queets), structurally complex "Olympic rainforests" contain massive Sitka spruce and western hemlock with an understory of

---

sect. *Bottom*: Percent cover of tree species in 639 ecological plots in the Olympic National Forest (downloaded from Ecoshare.info). Plots were divided into southwest and northeast areas. Curves are a locally weighted running median in a 300-m window

Table 1.1 Tree species common on the Olympic Peninsula and their climatic limits across their entire geographic range

Taxon name	Common name	January mean temp.	July mean temp.
		Min. (10th perc.)	(90th Perc.) Max.
Conifers			
Abies amabilis	Pacific silver fir	−13.8 (−8.2)	(19.1) 22.1
Abies grandis	Grand fir	−14.4 (−8.6)	(18.2) 20.5
Abies lasiocarpa	Subalpine fir	−30.0 (−23.2)	(16.1) 22.7
Cupressus nootkatensis	Alaska yellow cedar	−10.3 (−6.5)	(15.9) 19.2
Juniperus communis	Common juniper	−32.5 (−28.8)	(17.8) 25.7
Picea engelmannii	Engelmann spruce	−18.7 (−13.0)	(17.3) 24.4
Picea sitchensis	Sitka spruce	−12.3 (−6.9)	(16.8) 19.0
Pinus albicaulis	Whitebark pine	−18.7 (−14.7)	(16.6) 22.4
Pinus contorta	Lodgepole pine	−29.3 (−22.3)	(16.5) 27.6
Pinus monticola	Western white pine	−16.5 (−11.0)	(18.2) 23.5
Pseudotsuga menziesii	Douglas-fir	−16.9 (−11.5)	(20.0) 28.9
Taxus brevifolia	Pacific yew	−14.4 (−8.1)	(19.2) 24.3
Thuja plicata	Western redcedar	−17.1 (−11.1)	(17.9) 19.8
Tsuga heterophylla	Western hemlock	−17.1 (−11.8)	(17.1) 19.1
Tsuga mertensiana	Mountain hemlock	−15.6 (−9.5)	(16.4) 20.9
Hardwoods			
Acer circinatum	Vine maple	−10.7 (−5.2)	(19.1) 28.3
Acer glabrum	Rocky mountain maple	−23.8 (−12.0)	(28.1) 29.2
Acer macrophyllum	Bigleaf maple	−8.4 (−2.5)	(27.6) 29.8
Alnus rubra	Red alder	−15.6 (−6.5)	(19.5) 23.5
Alnus viridis ssp. sinuata	Slide alder	−28.0 (−17.1)	(16.6) 28.3
Arbutus menziesii	Pacific madrone	−4.4 (1.1)	(28.7) 29.6
Populus balsamifera ssp. trichocarpa	Black cottonwood	NA	NA
Quercus garryana	Oregon white oak	−4.1 (−0.4)	(26.1) 26.5

Table 1.1 (continued)

Taxon name	Common name	January mean temp.	July mean temp.
		Min. (10th perc.)	(90th Perc.) Max.
Rhamnus purshiana	Cascara	−14.4 (−8.7)	(18.9) 20.6
Rhododendron macrophyllum	Pacific rhododendron	−8.2 (−0.8)	(27.2) 27.4
Olympic Peninsula: entire region		−5.6 (−3.4)	(17.0) 18.3

Climatic limits are from mapped distributions overlayed on a 25-km climate grid (Thompson et al. 1999). Values shown are the minimum and 10th percentile of January mean temperature, and the 90th percentile and maximum value of July mean temperature, within the range of each species. Both statistics are presented because the minimum and maximum values may be affected by false outliers on the distribution maps. The last row shows corresponding values for the entire Olympic Peninsula using data resampled to a 1-km grid (Daly et al. 2002). This comparison shows that nearly all tree species on the peninsula can tolerate both colder and warmer temperatures than occur on the peninsula

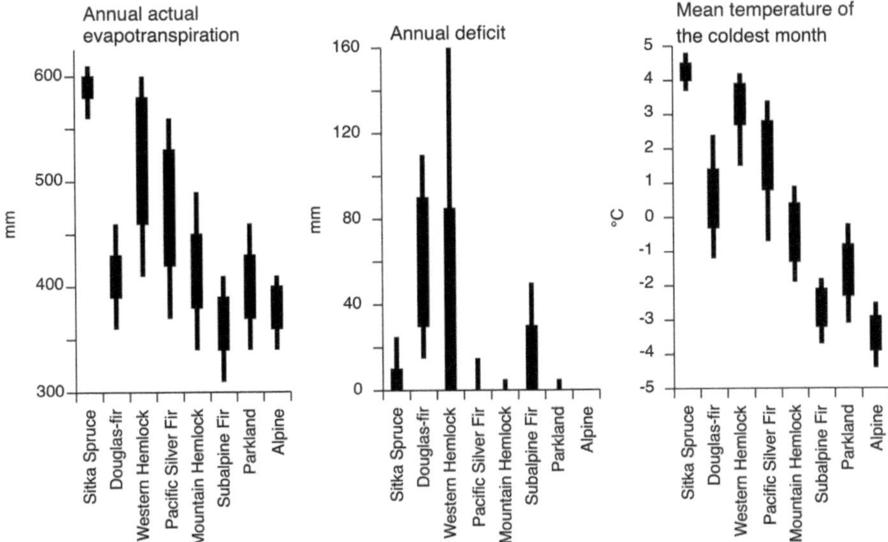

Fig. 1.11 Distribution of three bioclimatic variables within eight potential natural vegetation zones on the Olympic Peninsula. Box plots show the 10th–90th percentiles (*thin lines*) and the 25th–75th percentiles (*boxes*) of each variable within each vegetation zone. This figure was developed by overlaying the potential vegetation zones of Henderson et al. (2011) with a simple water balance model (Willmott et al. 1985; Gavin and Hu 2006) applied to mean monthly temperature and precipitation from the PRISM spatial climate data calculated for the 1970–1999 period at a grid resolution of 400 m (Daly et al. 2002)

epiphyte-laden bigleaf maple and open understory glades maintained by browsing Roosevelt elk (Schreiner et al. 1996). The forest contributes large woody debris to river channels that play critical roles in salmonid habitat and channel patterns. Stable logjams (>1000 years) result in complex floodplain surfaces and development of old forests on the highest protected surfaces (Montgomery and Abbe 2006). Forest structure is also affected by herbivory. An ungulate-exclosure study revealed that ungulate herbivory decreased herb, shrub, and tree density in the understory, and indicated that selective browsing may increase Sitka spruce recruitment relative to other conifers (Woodward et al. 1994). Second, on the coastal plain, a nutrient-poor muskeg swamp forest is dominated by western redcedar and shore pine with a very dense salal understory. Third, in several places, prairies dominated by bracken fern are thought to have been maintained by the burning practices of Native Americans (Reagan 1909; Anderson 2009). In many other places in the Sitka spruce zone, sphagnum bogs and fens maintain forest openings. These sites will be discussed again in Chap. 5.

The Sitka spruce zone has an average temperature of 4.3 °C in January, 15.6 °C in July, with 2700 mm of annual precipitation. This area is marked by strong fog, which results in up to 1 m of extra precipitation through fog drip and very little moisture deficit through the year (Fig. 1.11). These conditions favor Sitka spruce, which has low drought tolerance and is very tolerant of flooding disturbances

(Fig. 1.12; Peterson et al. 1997b; Niinemets and Valladares 2006). At smaller spatial scales, Sitka spruce is favored where topographic features, such as hollows and toe slopes, yield high soil moisture. Topographic effects are particularly important at its upper elevational limits, where Sitka spruce is limited to riparian areas (Fig. 1.10). During summer, a strong diurnal sea breeze and land breeze pattern moderates the daily high and low temperatures (Mass 1982).

Few tree-ring studies have been conducted within the Sitka spruce zone. One study showed that the relationship of Sitka spruce growth to climate is not straight-forward. Nakawatase and Peterson (2006) found that tree growth in this zone is poorly correlated with any climate variable, though there is a possibility that there is a correlation over multiyear frequencies. In contrast, Holman and Peterson (2006) found high interannual sensitivity in Sitka spruce growth rates, and a potential for very fast growth, and suggested that this forest type is the most sensitive to climate changes on the Olympic Peninsula. Considering the competitive environment of Sitka spruce forests, and the importance of canopy disturbances for determining competitive outcomes between Sitka spruce and more shade-tolerant species (Taylor 1990), it is likely that the effects of climate change in this forest zone may occur indirectly via changes in the disturbance regime.

1.4.2 *Western Hemlock Zone*

The western hemlock zone, occupying roughly half of the peninsula, occurs in a narrow elevational band above the Sitka spruce zone in the west, but occupies a broad band from sea level to nearly 1200 m in the drier northeast portion of the peninsula. Even though western hemlock may be common outside of this zone (Fig. 1.10), the western hemlock zone corresponds to areas where other species common at higher and lower elevation (e.g., Pacific silver fir and Sitka spruce) are rare or absent. In eastern and northern portion of this zone, forest fires have been common over the past several centuries, leading to dominance by even-aged seral Douglas-fir stands. Western hemlock and western redcedar are more common in wetter areas of the western peninsula, though pure climax forests are rare. Western white pine may occur sporadically, though this species has declined dramatically in recent decades (Harvey et al. 2008). Common shrubs include salal, vine maple, Oregon grape, red huckleberry, and salmonberry. Herbaceous species include swordfern, deerfern, oxalis, beargrass, twinflower, prince's pine, evergreen violet, vanillaleaf, trillium, and foamflower. The early-successional patterns within this zone are highly variable due to site differences and seed sources. On dry sites, Douglas-fir can invade quickly and produce even-aged stands within 10 years of the disturbance (Winter et al. 2002), while in wet sites, fireweed and shrub phases and/or a red alder phase may persist for decades. In addition, there is a broad continuum of variants of the western hemlock zone related to site moisture and productivity, each with well-defined understory indicator species (see Henderson et al. 1989).

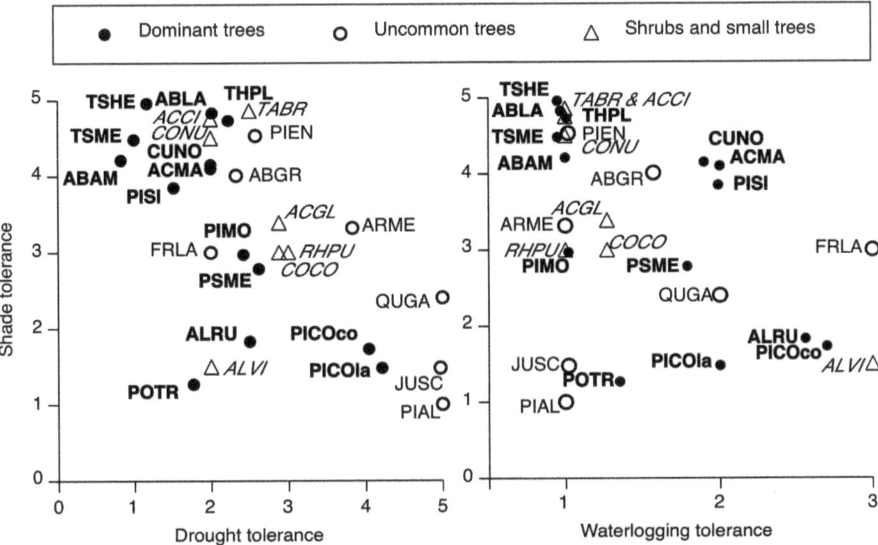

Fig. 1.12 Relative rankings of shade tolerance and drought tolerance of trees and selected shrubs on the Olympic Peninsula as reported in Niinemets and Valladares (2006). These characteristics explain a large amount of the geographic variation in species abundances. Low shade tolerance is associated with disturbed habitats (e.g., after fire, blowdowns, or along riparian corridors) and low drought tolerance is associated with the wettest climates on the peninsula, and low drought and shade tolerance occur in species restricted to wet disturbed sites (avalanche paths and along rivers). Tolerance to cold and length of growing season, not shown here, also varies considerably among species and explains much elevational variation. Species codes listed alphabetically: *ABAM* Pacific silver fir, *ABLA* subalpine fir, *ABGR* grand fir, *ACCI* vine maple *ACGL* Rocky Mountain maple, *ACMA* bigleaf maple, *ALRU* red alder, *ALVI* slide alder, *ARME* madrone, *COCO* beaked hazel, *CONU* Pacific dogwood, *CUNO* Alaska yellow cedar, *FRLA* Oregon white ash, *JUSC* Rocky Mountain juniper, *PIMO* western white pine, *PISI* Sitka spruce, *PICOco* shore pine, *PICOla* lodgepole pine, *PIAL* whitebark pine, *PIEN* Engelmann spruce, *POTR* black cottonwood, *PSME* Douglas-fir, *QUGA* garry oak, *RHPU* cascara, *TABR* Pacific yew, *THPL* western redcedar, *TSHE* western hemlock, *TSME* mountain hemlock

The western hemlock zone has an average temperature of 3.7 °C in January, 6.2 °C in July, with 2200 mm of annual precipitation. On the western peninsula, where western hemlock is most common, it is favored both by its very high shade tolerance in forests with little fire disturbance (Fig. 1.12) and by its requirement for high evapotranspiration (i.e., warm and wet summers; Gavin and Hu 2006). In dry northeastern areas, current forests of this zone are dominated by Douglas-fir, which spans a wide range of elevations due in part to recent widespread fires. Douglas-fir growth is limited by moisture deficits, except in the wettest portions of the peninsula (Littell et al. 2008), where climate and disturbance regimes favor western hemlock. Western redcedar is commonly associated with western hemlock, and is favored on the warmest sites with high growing season precipitation (>450 mm) or topographic moisture (Harrington and Gould 2010). Our calculation of moisture deficit in the western hemlock zone show a wide range of deficit values, but about 50% of the zone has no deficit (Fig. 1.11). Tree-ring studies have shown poor climatic control on the elevational boundaries of the dominant species in this zone, suggesting that competitive interactions may be a major determinant of species elevational boundaries (Ettinger et al. 2011).

1.4.3　Pacific Silver Fir Zone

The Pacific silver fir zone occurs at mid elevations, above the western hemlock zone and below the mountain hemlock zone. It may reach down to 300 m elevation on north-facing slopes on the western peninsula, but is restricted to areas >1000 m elevation in the drier eastern peninsula. The upper elevation limit of the zone is also much higher in the east (1400 m) than the west (1000 m). The dominant tree species are Pacific silver fir, western redcedar, and western hemlock, though Douglas-fir may be present up to 1000 m elevation as relicts from fires occurring during warmer periods 500–800 years ago. Common shrubs include several huckleberry species, white rhododendron, mountain ash, false huckleberry, salal, and Oregon grape. Herbaceous species include bunchberry, queen's cup, twisted stalk, vanilla-leaf, false lily-of-the-valley, deerfern, swordfern, five-leaved bramble, foamflower, and trillium. Succession after disturbance may begin with a mixture of western hemlock and Douglas-fir, which eventually are replaced by the very shade-tolerant and snow-load-tolerant Pacific silver fir.

There are few distinct variants within the Pacific silver fir zone. Slide alder communities form where snow avalanches recur frequently, or where creeping snow and high water tables inhibit seedling establishment. Alaska yellow cedar may be the only conifer that survives the conditions in these sites. Over broader areas, wet treeless meadows may be dominated by a thick cover of bracken fern and thimbleberry. These meadows appear to be compositionally stable and are not slowly converting to forest.

The Pacific silver fir zone has an average temperature of 2.0 °C in January, 15.1 °C in July, with 3500 mm of annual precipitation. Pacific silver fir is one of the least drought-tolerant trees on the peninsula (Fig. 1.12). Henderson et al. (2011) found that Pacific silver fir became common when it encountered areas of high precipitation, cold air drainages, north-facing aspects, and high topographic moisture as elevation increased. While actual evapotranspiration in the Pacific silver fir zone may be broadly similar with the western hemlock zone, moisture deficits and winter temperatures are much lower in the Pacific silver fir zone than at lower elevations (Fig. 1.11). Within this zone, snowpack may accumulate for a portion of the year, favoring Pacific silver fir, which has greater mechanical strength and flexibility under snow loads compared to western hemlock (note sapling in Fig. 1.9c).

1.4.4 Mountain Hemlock Zone

The mountain hemlock zone is the highest forested zone in the wettest portion of the peninsula, where it occurs from 1200 to 1500 m elevation. It is also present in patches on north-facing slopes in the northeast, but does not occur in dry high-elevation areas. In heavy-snow areas, subalpine parklands occur within the upper portions of the zone, while in low-snow areas moisture limits the occurrence of mountain hemlock and favors subalpine fir. The zone is dominated by both mountain hemlock and Pacific silver fir, with minor amounts of Alaska yellow cedar. Common shrubs include several huckleberry species, false huckleberry, mountain ash, white rhododendron, and red heather. Herbaceous species include avalanche lily, five-leaved bramble, deerfern, queen's cup, beargrass, and one-sided pyrola. Agee and Smith (1984) found that after fires, tree establishment generally mirrors that of the preexisting forest. Fire may promote huckleberry species in meadows. There are few variants of the mountain hemlock zone, though non-forested vegetation is relatively common.

The mountain hemlock zone has an average temperature of $-0.2\,°C$ in January, $13.1\,°C$ in July, with 3500 mm of annual precipitation. Moisture deficits are absent in this zone, and actual evapotranspiration and winter temperatures are notably lower than in the Pacific silver fir zone (Fig. 1.11). The upper elevational limit of the mountain hemlock zone is most likely limited by growing season length. A heavy snowpack protects mountain hemlock roots from freezing, and can provide moisture during the first half of the growing season (Means 1990). A heavy snowpack, however, can limit the radial growth of mountain hemlock by delaying the start of the growing season, resulting in high growth being associated with warm summer temperature (Gedalof and Smith 2001; Nakawatase and Peterson 2006). Climate strongly influences the growth of mountain hemlock adults and thus likely controls the upper elevational limit of trees in this zone as strong growth–climate relationship are found at tree line. In contrast, no such relationship exist at the lower elevational limit, suggesting climatic control of other life history stages (e.g., germination and establishment) and/or non-climatic factors (Ettinger et al. 2011) such as competition with tree species more common at lower elevations.

1.4.5 Subalpine Fir Zone

The subalpine fir zone occurs in the dry northeastern peninsula generally between 1300 and 1800 m elevation, and is the highest forest zone in the rain shadow climate. Depending on aspect and location, its lower boundary may abut the western hemlock, mountain hemlock, Douglas-Fir, or Pacific silver fir zones. Its upper boundary grades into subalpine parkland. The zone is dominated by subalpine fir and/or lodgepole pine, but may contain Douglas-fir. Shrubs include big huckleberry, common juniper, white rhododendron, and pachistima. Herbaceous plants include subalpine lupine, Sitka valerian, one-sided pyrola, Martindale's lomatium, and white hawkweed. Fire is the major disturbance in this region, and restocking of trees after fire may require more than five decades due to limited seedling recruitment from limitations of climate, substrate, and seed source (Agee and Smith 1984). Seedling survivorship is low in most meadows in the parkland zone, though heath-shrub communities tend to have safe sites for seedling survival (Soll 1994).

This zone has an average temperature of $-2.6\,°C$ in January, $12.2\,°C$ in July, with 2300 mm of annual precipitation. Zonal forests are most prevalent on dry south-facing slopes. The zone has a distinct moisture deficit approaching 50 mm (based on the simple water balance model in Fig. 1.11) and drought-adapted lodgepole pine may be abundant after fire. This zone experiences distinctly lower winter temperatures than in the mountain hemlock zone, its tree-line counterpart in the western peninsula (Fig. 1.11). Subalpine fir, like mountain hemlock, has high growth rates in years with low snowpack and high temperatures, and this relationship is especially strong at high-snowpack sites at highest elevations (Ettl and Peterson 1995; Peterson 1998). Growth is negatively correlated with temperature of the previous year, possibly via the impact of cone production and respiration on carbohydrate reserves. Furthermore, growth is correlated with summer precipitation at the middle and lower elevation ranges of subalpine fir, indicating sensitivity to drought varies elevationally (Ettl and Peterson 1995).

1.4.6 Douglas-fir Zone

The Douglas-fir zone is the smallest forest zone on the peninsula, located near the driest portions of the western hemlock zone on south- and west-facing slopes. It is observed in the Dungeness, Maiden Creek, and Elwha drainages. Douglas-fir is the dominant tree species, and grand fir, lodgepole pine, and madrone may be present. Conditions are too dry for western hemlock to be the late-successional dominant species (Fonda and Bliss 1969). Shrubs include kinnikinnick, Oregon grape, serviceberry, oceanspray, baldhip rose, snowberry, and salal. Herbaceous species present include western fescue, vanillaleaf, white hawksweed, prince's pine, Scouler's harebell, bigleaf sandwort, and starflower. Succession following disturbance is normally dominated by Douglas-fir during all stages. One distinct variant is the very dry kinnikinnick series, on shallow rocky soils with low fertility and frequent fire.

The Douglas-fir zone has an average temperature of $0.5\,°C$ in January, $15.0\,°C$ in July, with 1900 mm annual precipitation. Much of the area is composed of south-facing aspects and locations with low topographic moisture, making it the most consistently dry zone, with deficits between 20 and 100 mm. The combination of lower rainfall and cooler temperatures also limits the actual evapotranspiration to an average of 410 mm.

1.4.7   Subalpine Parkland Zone

The subalpine parkland zone occurs throughout high elevations on the peninsula. This zone is composed of scattered trees, meadows, and barrens. Its lower boundary is as low as 1100 m on north-facing slopes in wet areas and 1700 m in the dry northeast. The broad range of temperature and moisture in this zone results in several well-defined meadow community types. The first large survey by Kuramoto and Bliss (1970) revealed eight community types. Within meadows, consistent snowmelt gradients and topographic moisture also produce pronounced gradients in community types over short distances (Belsky and del Moral 1982). Further sampling by Schreiner (1994) expanded the eight communities of Kuramoto and Bliss (1970) to a system of 16 community types that cluster into four groups: scree and talus communities, *Phlox diffusa* communities, *Carex spectabilis* communities, and *Luetkea-Saxifraga* communities. Description of these communities is beyond the scope of this book.

The subalpine parkland zone has an average temperature of $-1.5\,°C$ in January, $12.1\,°C$ in July, with 2300 mm annual precipitation. The boundary between forested and parkland zones is likely highly dynamic and influenced by climate and fire, which is followed by slow tree regeneration. Invasion of dry meadows by subalpine fir requires heath-shrub microsites and normal or above-average growing-season moisture, such as conditions occurring when snowpack provided moisture to the soil during a high-snowpack period from 1956 to 1980 (Agee and Smith 1984; Woodward et al. 1995). In contrast, colonization in wet meadows occurs during a dry climate when snowpack does not limit the growing season, as occurred during 1921–1945 (Agee and Smith 1984; Woodward et al. 1995). Repeat photography of the parkland zone has also revealed meadow invasion by conifers as well as impacts from mountain goats (Schreiner 1994).

1.5   Natural Disturbance Regimes of the Olympic Peninsula

1.5.1   Fire

But the most magnificent thing I found, and to me it was an amazing discovery, was that every part of the Reserve I saw appeared to have been cleared by fire within the last few centuries. The mineral soil under the humus, wherever it was exposed about the roots of windfalls, was overlayed by a layer of charcoal and ashes. Continuous stretches of miles without a break were covered with a uniform growth of Douglas-fir from two to five feet in diameter, entirely unscarred by fire. Among them numerous rotting stumps of much larger trees did bear the marks of burning. I did not see a single young seedling of Douglas-fir under the forest cover, not a single opening made by fire which did not contain them. Fires conditioned and controlled the forest of the Olympics.
Gifford Pinchot, remarking on his 1897 visit to the Bogachiel area. *Breaking New Ground* (Pinchot 1947)

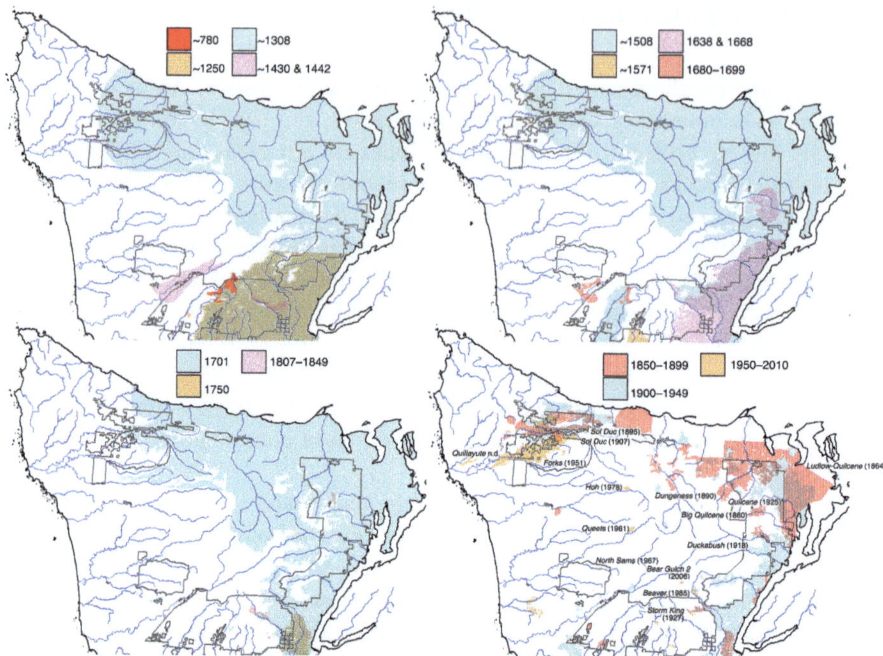

Fig. 1.13 Mapped fire events on the Olympic Peninsula compiled from historical records and Douglas-fir stand ages using data from Henderson and Peter (1981) and Henderson et al. (1989). See these sources for more detailed narratives of more severe fire years. This fire history focused on fires occurring within the Olympic National Forest (*gray* outline on maps) with an emphasis in the southeastern region. The extents of some fires are crude extrapolations and in some cases are truncated because fires extended beyond the focal area for particular studies (e.g., the AD 1250 fire). Most fires within the park are not shown except for approximations for the locations of the Queets and Hoh fires. Of the pre-1860 fires, most fire dates are estimates and fire extents are rough extrapolations, though the 1701 fire is accurately dated in several locations (Jan Henderson, *personal communication*)

Forest fire, even when recurring only every few centuries, profoundly affects the composition and structure of modern forests. Agee (1993) presented a rigorous synthesis of the fire ecology on the peninsula and Henderson (Henderson and Peter 1981; Henderson et al. 1989) synthesized evidence of fire history from various sources (Fig. 1.13). The following brief review of the above studies also includes references to more recent studies.

On the western Olympic Peninsula, the recent historical record suggests that fire is rare and of limited extent, although old-growth forests bear evidence of extensive fire in past centuries. In particular, Douglas-fir is mostly limited to areas that burned within the past 500 years. This conclusion is based on the observation that in dense forested settings fire is required for Douglas-fir establishment and its long lifespan, which allows it to persist as evidence of fire for centuries. Fire is especially rare in the coastal lowlands, where historical records show few major fires during the past 100 years and where the ages of Douglas-fir confirm the rarity of fire. For ex-

Fig. 1.14 Aerial photograph of the 1961 Queets fire scar photographed 3 years after the fire. Note the high severity of the burn but also several small patches of surviving trees. (Photo from Olympic National Park)

ample, Huff (1995) reported 515-year-old Douglas-fir stands in the Hoh drainage. Fahnestock and Agee (1983) reported a fire cycle (a fire interval estimated from the age distribution of trees over a large area) of 1100 years in Sitka spruce forests of coastal areas, though the actual fire cycle may be longer because many tree ages dated to wind disturbance rather than fire (Agee 1993). Agee and Flewelling (1983) used a probabilistic fire occurrence model based on modern climate and lightning occurrence and extrapolated a "natural fire rotation" of more than 4000 years for the western Olympic National Park. The natural fire rotation is the amount of time required to burn an area equal to the size of one's study area, and is conceptually similar to the fire cycle but is normally calculated from simple statistics of area burned per year (Heinselman 1973). For the western hemlock zone, a natural fire rotation estimate of 4000 years (calculated from recent rates of burning) conflicts with the evidence of widespread fire, hundreds of years ago (Huff 1995). The fire regime is better described as episodic, with centuries of little or no fire punctuated by years of regionally synchronous fire (Weisberg and Swanson 2003).

Fire is more common at mid elevations, above river terraces, than in lowlands because south-facing aspects are dry during the late summer months. A fire on the western Olympic Mountains, the 96-ha Queets fire in 1961, appears to be typical of lightning fires in similar areas as inferred from stand ages (Fig. 1.14). This was a high-severity fire, but was restricted to steep south-facing slopes and did not spread into terrace and riparian forests or into upper-elevation forests with late snowmelt date or more than a few kilometers cross-slope. Schmidt (1960) claimed that this spatial pattern of fire was the norm in similar forests on Vancouver Island, and an intensive study of soil charcoal there, also showed that the vast majority of fire over the Holocene was restricted spatially along slopes similar to the Queets fire (Gavin et al. 2003a). Extreme fire weather, however, can overcome these topographic barriers. The 425-ha Hoh fire in 1978, ignited by lightning after nearly 3 months of no rainfall, extended into both riparian and subalpine forests on a south-facing slope of the upper Hoh River.

In striking contrast to west-side forests, the east-side western hemlock zone forests have been shaped by fire over the past few 100 years. Indeed, the ages of Douglas-fir trees suggest extensive fires in ca. 1308, ca. 1508, and 1701 (Fig. 1.13). Extensive fires also occurred during settlement times (1860–1910), especially in the northeastern portion of the peninsula and along the Hood Canal. These fires are almost all due to land-clearance activities and escaped fires from logging operations. Though not in eastern areas, several fires in the Lake Constance area (Sol Duc fires of 1895 and 1907 and the Forks fire of 1951) had explosive crown-fire runs resulting from strong east winds (Agee 1993). Such winds occur in the Strait of Juan de Fuca when a thermal trough develops over the Pacific Ocean (Brewer et al. 2012). Most major fires on the peninsula occur during these east wind events (Agee 1993). Another important consideration is that fire activity during the past 50 years is a small fraction of the area burned during the early twentieth century (Fig. 1.15), likely due to increased suppression activity and a reduction in human-caused ignitions.

In addition to the dataset of mapped major fire events on the peninsula, Pickford et al. (1980) presented a thorough inventory of 747 fire events within the Olympic National Park from 1916 to 1975 (Fig. 1.16). Of the 274 fires that were lightning caused, the majority occurred in only 6 years. The proportion of human-caused fires decreased dramatically since 1946 (burning a total of only 10 ha), and the great majority of lightning fire events were small fires. Indeed, only 10 fires greater than 30 ha in size could account for 76 % of the area burned. In the driest portion of the park, for example, very steep rocky terrain results in discontinuous fuel and a heterogeneous pattern of burn severity, as occurred in the 2009 Heat Wave complex. Thus, the natural pattern of fire is not only limited by fuel moisture and ignition but also by the way topography interrupts fuel continuity.

These aforementioned controls of fire spread result in the observed distribution of fire sizes over time. The frequency distribution of fire sizes thus is an emergent property of the many factors that control fire behavior and spread. The cumulative fire size distribution provides evidence of the controls of topography and weather and fuels; linear relationships suggest agreement with predictions of fractal behavior of fire (Moritz et al. 2005). For the 353 lightning fires on the peninsula since 1900, the cumulative frequency-area distribution shows two distinct breakpoints (Fig. 1.15). Such breakpoints may point to ecological factors that affect sets of fires similarly (Ricotta et al. 2001). Specifically, only 12 % of the fires were recorded at sizes under 0.04 km^2 (4 ha) possibly because once a fire is ignited and has a potential to spread, it will frequently spread to sizes greater than 4 ha. It is also possible that many such small fires were undetected and are not in the fire history database. The majority of fire (72 %) occurred in the range of 0.04 and 4 km^2. For the 16 % of fires larger than 4 km^2, the frequency-area distribution suggests that fire size increased at a more constrained rate, never extending above 100 km^2. Topographic and fuel barriers such as ridges, rivers, and climatic gradients are likely imposing

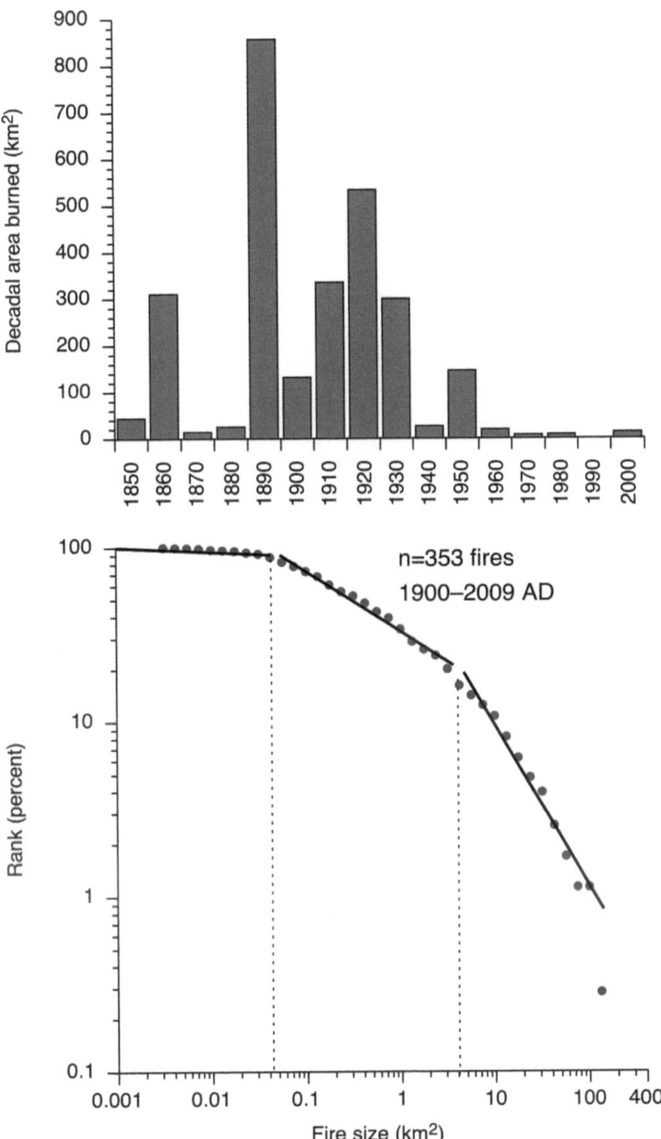

Fig. 1.15 Top: Area burned in decadal time steps from 1850 to present from the mapped fires presented in Fig. 1.11 and from the Monitoring Trends in Burn Severity project (Eidenshink et al. 2007). Bottom: Inverse cumulative size distribution of recorded fires on the Olympic Peninsula from 1900 to present. Most fires were small: 90 % of the 353 reported fires were smaller than 10 km^2

size constraints on fires within this size range in ways that were less important for smaller fires. Ricotta et al. (2001) found a similar pattern of breakpoints spanning two orders of magnitude of fire size. While there is a debate regarding the

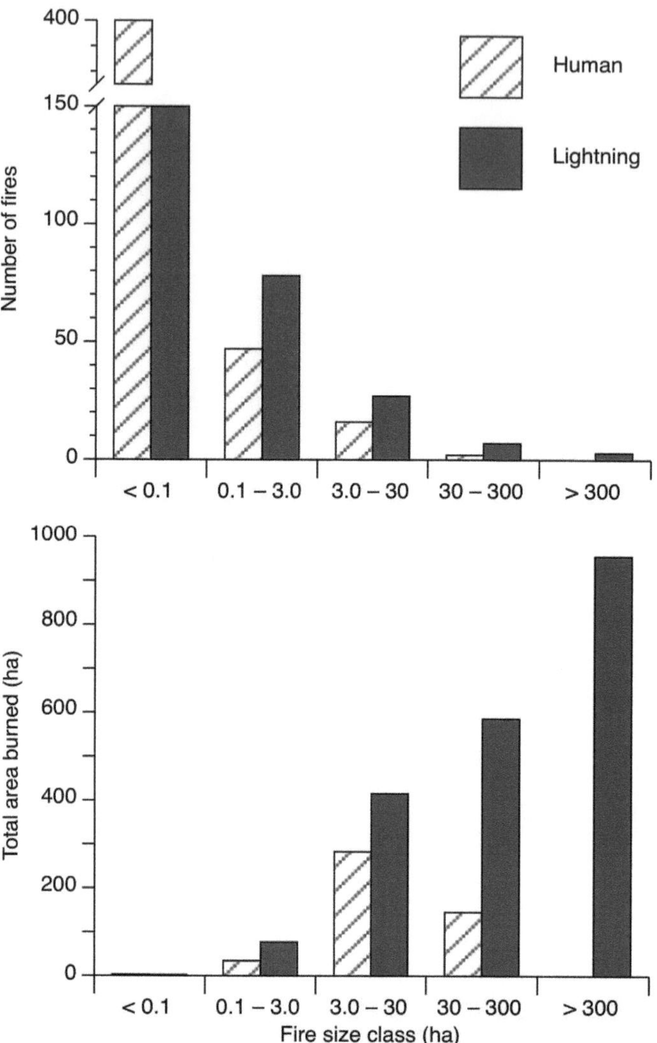

Fig. 1.16 Number of fires and area burned by fire in the Olympic National Park from 1916 to 1975 as reported in Pickford et al. (1980). Human- versus lightning-caused fires are plotted separately

application of complexity theory to the type of power–law relationship in Fig. 1.15, the cumulative frequency distribution remains a useful means of interpreting ecological processes controlling fire extent (Millington et al. 2006).

Tree-ring studies of fire, corroborate the episodic pattern of fire over centennial-scale time periods. In the Morse Creek drainage south of Port Angeles, Douglas-fir trees reveal growth releases corresponding to two major fire periods at ca. 1700 and 1895, in addition to many other fires of smaller extent (Wetzel and Fonda 2000). Though criteria for identifying fires from tree-rings were less rigorous than for most other such studies (i.e., from growth releases rather than by unambiguous fire scars), this study revealed that at a scale of 200 ha, fire recurs at an average of 21 years. In the Elwha River south of Lake Mills, another tree-ring fire scar study mapped fire events for the past 440 years (Wendel and Zabowski 2010). This study used conservative criteria for identifying fires (fires defined by three scars), which resulted in longer fire interval estimates than found by Wetzel and Fonda (2000) in a similar forest type. The fire rotation was 146 years prior to 1775, when large fires burned in 1568, 1661, 1676, and 1729. No fires were detected from 1775 to 1850 due to a combination of the effects of cooler Little Ice Age and the dramatic decline in Native American population. During the period of Euro-American settler influx, the fire rotation was calculated to be 53 years. Wendel and Zabowski (2010) suggested that the early period of high fire was partly due to a feedback involving fire-induced loading of fine fuels causing reburns (Agee and Huff 1987), which caused the large fires detected in 1868, 1890, and 1898. Few fires burned after 1926.

The effective exclusion of fire since the 1940s has resulted in increased density of understory vegetation and a simplification of forest structure relative to what occurred during earlier centuries with more frequent fire (Weisberg 2004). As in other forests in the Pacific Northwest, the occurrence of this forest structure in the Douglas-fir zone has prompted the application of prescribed fire to reduce stocking of small-diameter trees and potential for high-severity fire. Initial results are promising as the recent application of a low-intensity prescribed fire in the Maiden Creek watershed has been successful in reducing understory biomass while incurring little canopy tree mortality (Fonda and Binney 2011).

1.5.2 *Wind*

Severe winter cyclonic storms can produce winds that approach or surpass hurricane forces (74 mph) and lead to extensive forest blowdowns. Tree-ring studies along the Oregon coast have found that wind disturbances have been the primary control on tree growth rates, with the growth-release evidence of high-wind events most common in the first half of the twentieth century (Knapp and Hadley 2012).

The three largest recorded wind events on the Olympic Peninsula occurred in 1921 (The 21 Blow), 1962 (The Columbus Day Storm), and 1997 (Fig. 1.17). The 1921 storm was reported to have reached wind speeds more than 100 mph along the coast resulting in 40 % of the forest area blown down on the southwestern flanks of the Olympic Mountains (Mass and Dotson 2010). Windthrow from the 21 Blow was reported to have been strongly biased to western hemlock and Pacific silver fir, as those species have weaker boles compared to western redcedar and Douglas-fir. The blowdown also resulted in a release of the advance regeneration in the forests, understory trees which were predominately the shade-tolerant western hemlock (Dale et al. 1986). The Columbus Day Storm originated as a tropical cyclone in the western Pacific that was entrained into the westerlies and moved quickly to the Pacific Northwest (Lynott and Cramer 1966). The majority of damage was focused in the southern peninsula. Another storm in 1997 had winds focused along the coast with

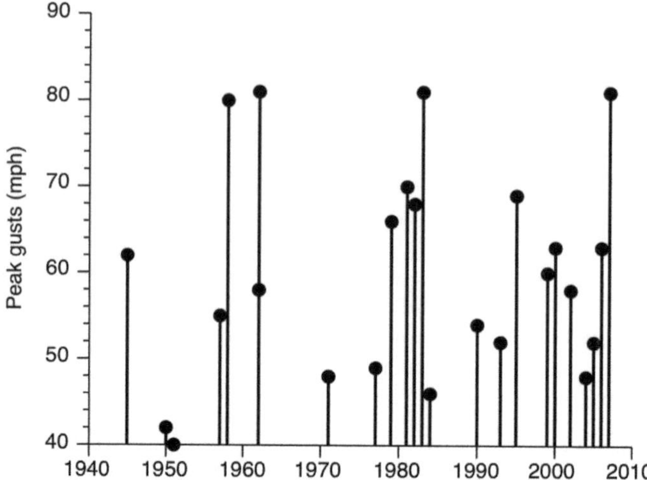

Fig. 1.17 Peak gusts at Hoquiam WA of 24 storms between 1945 and 2010, as reported on the Washington State Climatologist web page (www.climate.washington.edu/stormking). Wind gusts of ca. 80 mph recur at approximately 20-year intervals

most blowdown occurring south and north of Grays Harbor. Within a 30-mile-wide coastal strip, 11,600 ha of blowdown was mapped, much of it occurring on tribal land (Forest Health Program 2008).

1.5.3   Other Disturbance Agents

Numerous other disturbance types undoubtedly affect forests of the Olympic Peninsula. However, their occurrence is less well documented. This section reviews other processes of generating landscape disturbances, including insects and pathogens, herbivory, and geomorphic disturbances.

With the caveat that few insect and disease surveys have been conducted within the park, few widespread outbreaks have been reported and, at least until recently, forest insects are considered to play a minor role in the disturbance regime (Henderson et al. 1989). Although few insects have been ranked as a severe threat (Devine et al. 2012), potential insect damage has been highlighted in recent reports. Balsam wooly adelgid (*Adelges piceae*), a nonnative defoliating insect that attacks subalpine fir and potentially other true fir species, is considered a major threat into the future (Devine et al. 2012). Preliminary results of a survey indicate that high levels of defoliation are predisposed by high stand density and warm winter temperatures, suggesting increasing outbreaks with increasing winter temperature (Hutten et al. 2012). Western balsam bark beetles (*Dryocoetes confusus*) may also pose a threat for high-elevation subalpine fir, as weakened trees are more susceptible to the beetles. The hemlock looper (*Lambdina fiscellaria*), a defoliating lepidopteran insect, may have outbreaks on western hemlock causing high levels of defoliation within restricted patches. Outbreaks are initiated following periods of high temperature and low precipitation, which may become more common in the future (McCloskey et al. 2009). Other forest insects may be similarly favored by such climate changes (Lysak et al. 2006).

The dramatic reduction of western white pine constitutes another new unique disturbance within the park. Western white pine may have been more common throughout the peninsula in the recent past, but has suffered mortality from white pine blister rust (*Cronartium ribicola*) which was accidentally introduced near Vancouver BC in 1910. Western white pine was detected in only 5 % of the 697 forest ecology plots on the peninsula (Fig. 1.10), which likely reflects a reduction from an unknown level, prior to the blister rust. Whitebark pine, the other five-needle pine (subgenus Strobus) in the region that is susceptible to the blister rust, is widely scattered in small clumps at high elevation in the eastern and northeastern peninsula. Studies have shown an infection rate of about 20 % (Ward et al. 2006). It is possible that the highly discontinuous distribution of whitebark pine may be slowing the spread of blister rust.

Herbivory by ungulates, especially elk, constitutes another form of disturbance, although it is unclear what level of grazing has occurred in coastal forests over long periods (e.g., hundreds of years). Elk grazing creates glades in the Sitka spruce zone by limiting tree recruitment, especially maple and cottonwood in riparian areas. The extirpation of wolves from the peninsula in the early 1900s has resulted in a rapid increase in elk and reduced recruitment of riparian trees (Beschta and Ripple 2008). The response of tree-seedling abundance to elk densities, however, requires further investigation, as study on western hemlock ages failed to show a relationship to the changing elk populations (Harmon and Franklin 1983). Unusually high levels of herbivory, as a form of punctuated modern disturbance for plant populations, have interesting implications. For example, the decline of riparian tree regeneration may reduce bank stability and result in a widening of the active channel, exposing bare alluvium, and reducing the flows of biota and nutrients between riparian forests and the active river channel (Beschta and Ripple 2008).

The mountain goat (*Oreamnos americanus*) was introduced to the peninsula in the 1920s. The population grew rapidly to more than 1000 animals by the early 1980s. Concern for damage to meadows and endemic plant species rose after photographic and plot data indicated major meadow loss from herbivory, trampling, and wallowing (Schreiner et al. 1994; Schreiner and Burger 1994). This prompted a plan to remove goats, which reduced the population to fewer than 300 by 1994. After the goal of complete removal was abandoned, the population has begun to grow at about 4 % per year (Jenkins et al. 2012b). Lyman (1995) has questioned the strength of evidence supporting the goat's nonnative status, as the fossil record within the mountainous part of the peninsula is very limited. For example, on Vancouver Island, where karst caves provide the conditions for bone preservation, mountain goats were present during the late-glacial period but are absent today (Nagorsen and Keddie 2000). The management of mountain goats in the park remains controversial.

Other native herbivores produce important disturbances. The Olympic marmot disturbs meadows by selective herbivory and burrows that create large mounds of soil. Moderate levels of marmot activity may reduce the abundance of the most common species and thereby increase plant diversity. However, where marmot populations are high, palatable species are greatly reduced and ruderal (weedy) species increase (del Moral 1984). Black bear damage to young forests is very widespread on the Olympic Peninsula. One bear may girdle 60–70 trees per day while feeding on phloem, with preference for trees 20–40 cm in diameter (Ziegltrum 2004). Logging and replanting outside the park has created extensive stands of trees suitable for such feeding and bear damage has become an economic and wildlife management issue.

Geomorphic disturbances are frequent due to the steep terrain within the park, the sedimentary bedrock, and the high rain and snowfall. An inventory of landslides by Gerstel (1999) based on aerial photographs over approximately half of the western peninsula outside of the park, identified 465 deep-seated (below rooting depth) landslides. Another database of mapped landslides shows that 3.8 % of the mountainous landscape of the south-central portion of the park has the signature of a landslide disturbance (Fig. 1.18). The high precipitation on ridgetops is a major factor affecting the geographic distribution of landsliding (Minder et al. 2009). Indeed, the orientation of landsliding is biased to the south and southwest aspects (Fig. 1.18), which receive more rainfall than leeward aspects. The same orientation of landslides was noted in a similar setting on Vancouver Island (Jakob 2000).

Large landslides provide opportunities for primary succession in a landscape where other large disturbances are rare. High light levels and exposed mineral soils on fresh landslides are ideal for the establishment of Douglas-fir and pine species that otherwise require high-severity fire for establishment (Geertsema and Pojar 2007). Considering the aerial extent and frequency of natural landsliding, these events are likely crucial in maintaining a wide diversity of disturbance-adapted species particularly within western areas of the park where fires are uncommon (Swanson et al. 1988, 2010). For example, a landslide that dammed the Elwha River in 1967 was followed by a dam-break flood that produced a wide range of surface features on the outwash fan that promoted a range of successional trajectories (Acker et al. 2008). Landsliding is also an important source of sediment supply that shapes the geomorphology of floodplains and width and stability of active channels of major rivers; such characteristics vary greatly among the Queets and Quinault rivers (O'Connor et al. 2003).

Very large deep-seated paleo-landslides may have been triggered by major earthquakes. A series of rock avalanches in the southeastern Olympics all date to the period of 1000–1300 years ago, based on radiocarbon dating of drowned trees in landslide-dammed lakes (Schuster et al. 1992). A nearby site on Saddle Mountain yielded landslide ages of 1600 and 1250 years ago (Wilson et al. 1979). The older event and a much older event on the same fault were found in a more recent study (Witter et al. 2008). Another series of very large landslides from Storm King Mountain dammed the ancient Lake Crescent and changed its outlet to the Lyre River. While some of these landslides may have occurred within the past 500 years, the

Fig. 1.18 *Top*: Mapped landslides (yellow) over the upper Quinault River in the Olympic National Park. Recent landslides and older revegetated landslides are not differentiated. Elevations range from approximately 200–1600 m. Data are from the Washington State Landslide Database (http://www.dnr.wa.gov/ResearchScience/Pages/PubData.aspx). *Bottom*: The Grouse Creek landslide of 1997, which was triggered by a rain-on-snow event (Serder 1999). The run-out of the landslide is 2.4 km. (Photo by Karl Wegmann)

age of the largest landslides remain unknown (Logan and Schuster 1991). Both of these sites are on areas of basalt bedrock, which may require seismic activity to produce major landslides (Schuster et al. 1992).

The coast of the Olympic Peninsula has experienced repeated seismic events due to its proximity to the Cascadia subduction zone, in which the Juan de Fuca plate dips below the North American plate tens of kilometers offshore of Washington and Oregon. However, the geology of the peninsula itself seems to be particularly resistant to seismic activity. Only 14 earthquakes of magnitude 5.5 or greater have occurred near the peninsula since 1965 (http://earthquake.usgs.gov/earthquakes/eqarchives/). Only four other notable earthquakes occurred earlier in the century (Yeats 2004). Three other events far from Washington produced tsunamis with mi-

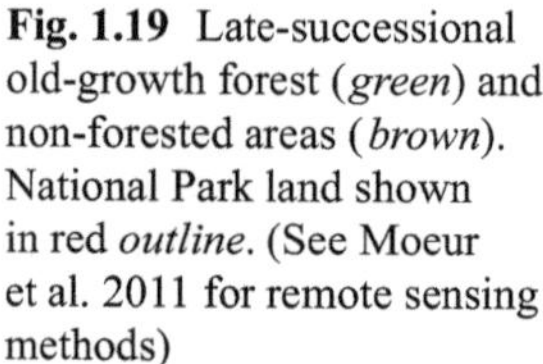

Fig. 1.19 Late-successional
old-growth forest (*green*) and
non-forested areas (*brown*).
National Park land shown
in red *outline*. (See Moeur
et al. 2011 for remote sensing
methods)

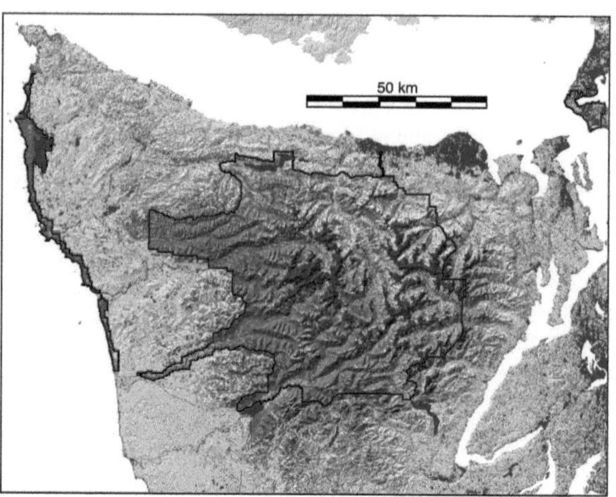

nor impacts on the peninsula (Yeats 2004). It is clear from the recent seismic record that the peninsula is at risk from activity along the Cascadia subduction zone, as half of the recent earthquakes were identified as "offshore."

Paleoseismic evidence for large subduction zone earthquakes is increasing. Atwater (1987) suggests that methods of interpreting stratigraphic sections can pinpoint locations of subduction zone earthquake related subsidence in coastal lowlying areas. For example, he concludes that the 1964 Seattle earthquake caused such subsidence, mostly in estuaries. Willapa Bay sediments recorded seven such events in the past 7000 years (Atwater 1987). In AD 1700, a major earthquake of magnitude 9 affected much of the Pacific Northwest southward from the southern peninsula (Atwater and Yamaguchi 1991; Atwater et al. 2005). The impact of these events on coastal villages is discussed in Chap. 5.

Today, forest harvesting constitutes the main form of disturbance on the peninsula. Currently, the disturbance regimes outside the park are likely very different from those within the park because extensive young stands and changed species composition in forests, replanted after harvest, alter the probability of disturbances compared to older forests within the park. The extent of the loss of old-growth forest is clear from an analysis of late-successional old-growth forest from Landsat imagery (Moeur et al. 2011; Fig. 1.19). The maps also show that old-growth forests are concentrated in the western half of the park. Historical fire disturbances reduced the late-successional forest area in eastern portions of the park.

Chapter 2
Geology and Historical Biogeography of the Olympic Peninsula

Abstract A brief introduction to the geologic and climatic history of the Olympic Peninsula is presented, focusing on the Middle Miocene to present (the past 20 million years). The period is marked by a general cooling and drying trend, with a corresponding loss of the mesic broadleaf forest and the rise of conifer-dominated forests. The causes of the regional climate changes are discussed with an eye on the relative roles of topographic uplift versus larger global-scale changes in atmospheric circulation and sea surface temperature. The patterns of endemism and disjunction on the peninsula are presented to emphasize insular aspects of its biogeography. A list of 29 endemic taxa is compiled from literature searches, though it is acknowledged the list will continue to grow. Mapped distributions of the endemic taxa suggest long-term persistence of certain habitat types, including dry meadows and wet riparian forests, through periods of glacial cycles.

The gap that separates the Olympic Mountains from the Cascade Range…is only about a hundred miles wide and is filled by the dense forest already mentioned, affording continuity of range…but the higher parts of the Olympics…have been disconnected from corresponding zones to the north and east for at least several million years, a period long enough to admit a considerable amount of differentiation in the species stranded here…
C. Hart Merriam, undated manuscript, quoted in Schultz (1994)

2.1 Late Cenozoic Geologic and Climatic History

During the Eocene, about 50 million years ago, a deep marine rift formed parallel to the continent as the oceanic (Kula) plate moved obliquely to the continental (North American) plate (Fig. 2.1). This rift caused immense amounts of lava to accumulate, forming a large region of basalts under the ocean (Babcock et al. 1992). Sediments accumulated on both the continental and oceanic side of the basalts. As the oceanic plate pushed against the continental plate, both the oceanic sediments and the basalt were compressed and turned upward. This process of subduction under the continental plate resulted in the accretion of a large wedge of sediments against the continent, which were pushed vertically and forced into an eastward-plunging arch (Tabor 1975). Younger sedimentary rocks were the last to be pushed against

© Springer International Publishing Switzerland 2015

D. G. Gavin, L. B. Brubaker, *Late Pleistocene and Holocene Environmental Change on the Olympic Peninsula, Washington,* Ecological Studies 222,
DOI 10.1007/978-3-319-11014-1_2

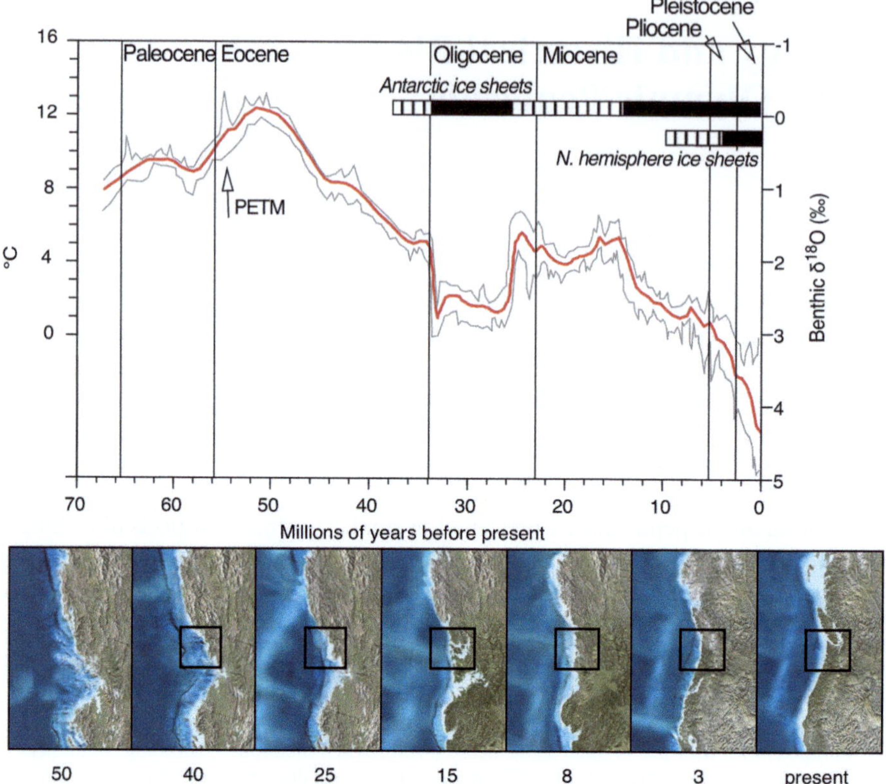

Fig. 2.1 *Top:* Global climate trajectory during the Cenozoic inferred from oxygen isotope ratios of benthic ocean sediment (modified from Zachos et al. 2001). *Bottom:* Paleogeography of western North America. Colors correspond to vegetation cover. (Maps from Ron Blakey, Colorado Plateau Geosystems, Inc, http://www.cpgeosystems.com/paleomaps.html)

the continent. Because the basalt mass, and the continent behind it, were much more rigid than the oceanic sediments, the marine sandstones, siltstones, and conglomerates making up the core of the Olympic Mountains were disrupted and folded. The horseshoe shape of the basalt Crescent Formation (Fig. 2.2) may have resulted from the ocean plate forcing the basalt mass into the "inside corner" between Vancouver Island and the continent. Beginning around 20 million years ago, the lighter sedimentary rocks rose relative to the dense oceanic crust leading to the uplift of the Olympic Mountains. Ages on the central core rocks of the peninsula, dated by the zircon fission track method, suggest the uplift of the mountains has been occurring for 17 million years and that the erosion of the terrain is in rough equilibrium with the uplift (Pazzaglia and Brandon 2001; Wegmann and Pazzaglia 2002). However, there remains much uncertainty about the age of the marine sedimentary rocks, which are often mapped within broad time periods (e.g., Miocene to Eocene).

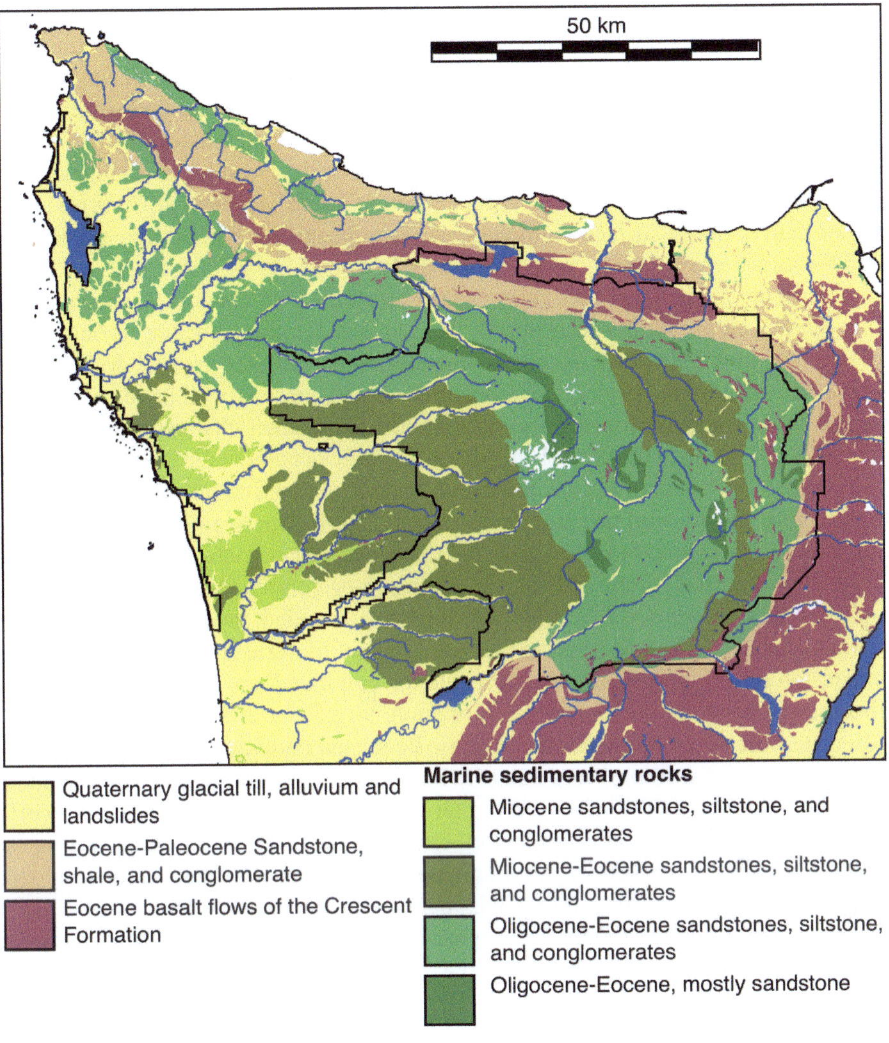

Fig. 2.2 Geological map of the Olympic Peninsula showing major units differentiated by age class. (Data from Washington Division of Geology and Earth Resources 2008)

During the uplift of the Olympic Mountains, from the Middle Miocene to present, the Olympic Mountains experienced large climatic changes. A warming in the Middle Miocene from 18 to 16 Ma, the Miocene Climatic Optimum, interrupted a general global cooling trend during the Cenozoic. Vegetation at this time in the Pacific Northwest indicates summer moisture and humidity sufficient to maintain a diversity of broadleaf deciduous trees. The warm sea surface temperatures of the Miocene fed summer moisture into western North America in what was likely a "Rocky Mountain Monsoon" with probably more than 400 mm of precipitation in the summer months alone (Lyle et al. 2003) and thus much stronger than the modest modern monsoon in Arizona (note increased vegetation cover at 15 Ma as shown in Fig. 2.1). It is difficult to simulate the warm and wet climate of the Middle Miocene under the CO_2 levels estimated for that time, which were similar to modern levels (<400 ppm; Goldner et al. 2013). During the late Miocene, the western USA began shifting from a summer-wet/winter-dry climate to a summer-dry/winter-wet climate. This gradual change, from 15 to 5 Ma, was likely linked to changes in sea surface temperature which would reduce summer rainfall (Lyle et al. 2008). The ultimate causes of sea surface temperatures, via atmospheric and oceanic circulation and the global energy balance, are poorly understood.

The uplift of the Cascade/Sierra ranges created a rain shadow to the east, but this uplift was probably not primarily responsible for the Miocene drying in the region. The mountain uplift occurred generally earlier than the gradual drying trend. More importantly, the drying trend occurred on the coast as well as in the modern rain shadow environment, suggesting a more fundamental change in atmospheric circulation (Lyle et al. 2008). The drying was likely accompanied by an increase in the strength of the Pacific subtropical high system and coastal upwelling, which are responsible for the summer aridity in western North America (Wolfe 1985). This climate change resulted in great losses of the deciduous element that required warm and moist summers, while promoting expansion of conifer forests. In addition, as uplift of the Cascade and Sierra Nevada Ranges proceeded, the interior rain shadow climate developed and resulted in the development of xeric steppe vegetation. The expansion of conifers and woodland vegetation continued into the Pliocene. Despite this major change in the vegetation formation, 90 genera of the Pacific Northwest Miocene flora occur today on the Olympic Peninsula (Buckingham et al. 1995).

Beginning 2.6 million years ago (the Pleistocene), global climate completed its transition from a "greenhouse" world that lacks ice sheets to an "ice-house" world marked by cyclic glaciations in the northern hemisphere. Glacial periods lasting ca. 90,000 years were marked by rapid climate changes and ice sheets extending over most of Canada. In contrast, interglacial periods lasting ca. 10,000 years resulted in species expanding their ranges into higher latitudes. The climate history of the late Pleistocene is the focus of Chaps. 3 and 4.

2.2 Endemism, Disjunction, and the Insular Nature of the Olympic Biota

The insular (island-like) nature of the plant and animal life of the Olympic Peninsula has been widely noted (Houston et al. 1994; McNulty 2009). Three main features of the biodiversity on the peninsula are consistent with long-term isolation and the effects of the mountains functioning as an ice-age refugium during the Pleistocene. First, 29 plant and animal taxa are known to be endemic to the Olympic Peninsula (Table 2.1, Fig. 2.3). Second, at least 13 species common in the Cascade Range were historically absent from the peninsula, suggesting the operation of a long-term

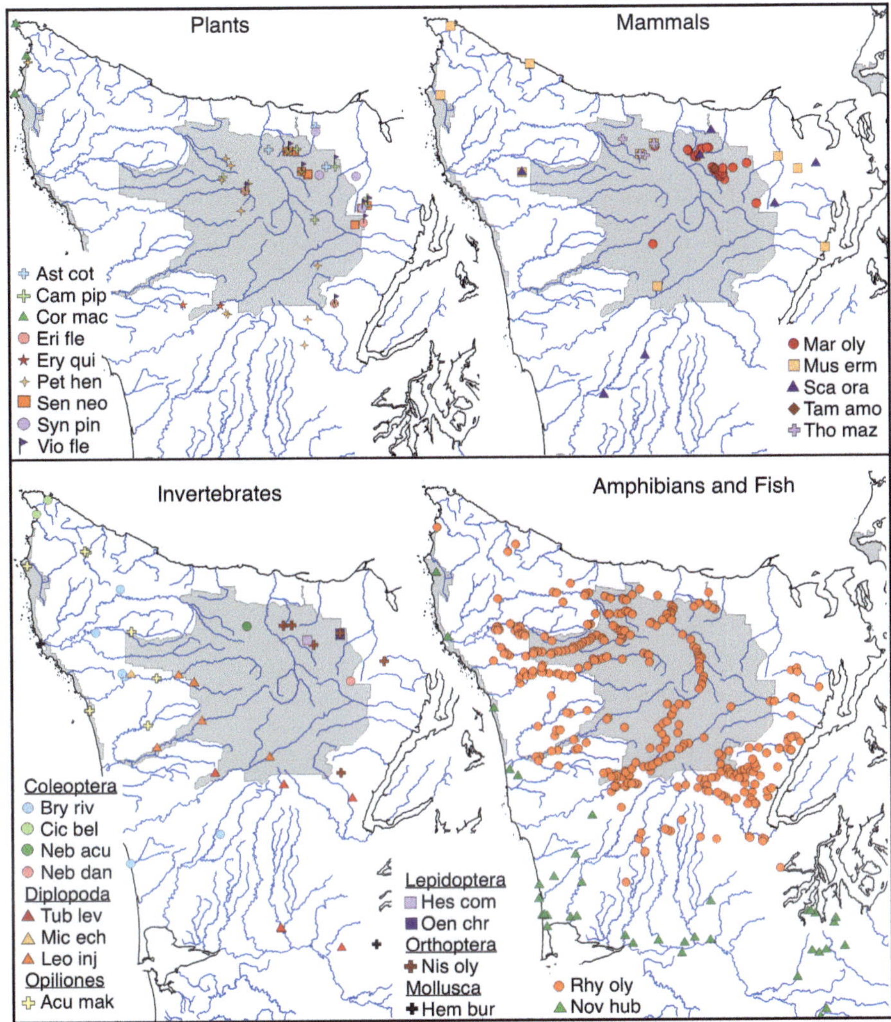

Fig. 2.3 Distribution of endemic (or near-endemic) taxa of the Olympic Peninsula. *Six-character species codes* are first characters of genus and species names listed in Table 2.1.

barrier to dispersal. Third, a large number of species on the peninsula are disjunct from their closest conspecifics or sister taxa by hundreds of kilometers.

The endemic taxa of the Olympic Peninsula are found in a small set of habitats, suggesting that these habitats are a long-term feature of the peninsula though not necessarily in their current extent and spatial distribution (Fig. 2.4). Of the 29 endemic taxa, 13 have ranges restricted to subalpine or higher elevations in the north or northeast. In particular, five species are associated with scree or rocky slopes. These dry meadow and woodland habitats are more extensive in the modern Rocky Mountains and may have been more common in the past on the Olympic Peninsula,

Table 2.1 Endemic taxa of the Olympic Peninsula. (This list was developed from Park Service web pages http://www.nps.gov/olym/naturescience/endemic-animals.htm, Houston et al. 1994, Buckingham et al. 1995, McNulty 2009, and several more recent studies. Plant taxonomy was checked against the flora of North America. Note that some taxa have ranges extending partly outside the peninsula)

Scientific name	Common name	Habitat association (and quadrant on Peninsula)	Source (for habitat association)
Vascular plants			
Astralagaus cottonii	Olympic Mountain milkvetch	Subalpine, open sites (northeastern mountains)	Buckingham et al. 1995
Campanula piperi	Piper's bellflower	Montane to alpine, rocky sites (Northeast to central)	Buckingham et al. 1995
Corallorhiza maculata var. *ozettensis*	Spotted coral root	Lowland, partial shade (Northwest)	Tisch 2001
Erigeron flettii	Flett's fleabane	Subalpine to alpine, open sites (Northeast to south)	Buckingham et al. 1995
Erigeron peregrinus ssp. *peregrinus* var. *thompsonii*	Thompson's wandering fleabane	Lowland, bog sites (Southwest)	Buckingham et al. 1995
Erythronium quinaultense	Quinault fawn lily	Lowland, open or partially open (Southwest)	Allen 2001
Petrophytum hendersonii	Olympic rock mat	Montane to alpine, rocky sites (South, north, and east)	Buckingham et al. 1995
Senecio neowebsteri	Olympic Mountain groundsel	Subalpine to alpine, scree sites (North, northeast, and central)	Buckingham et al. 1995
Synthyris pinnatifida var. *lanuginosa*	Olympic Mountain synthyris	Subalpine to alpine, scree sites (North, northeast, and central)	Buckingham et al. 1995
Viola flettii	Flett's violet	Subalpine to alpine, rocky sites (North, northeast, and central)	Buckingham et al. 1995
Amphibians			
Rhyacotriton olympicus	Olympic torrent salamander	Lowlands to montane, steep gradient streams (Southwest)	Adams and Bury 2002
Mammals			
Marmota olympus	Olympic marmot	Subalpine, open meadows (Throughout)	Edelman 2003
Tamias amoenus caurinus	Olympic yellow-pine chipmunk	Subalpine, forest, and parkland (North and northeast)	Sutton 1992; Demboski and Sullivan 2003
Scapanus orarius (Olympic clade)	coast mole	Subalpine (Northeast and southwest)	Welch 2008[a]
Thomomys mazama melanops	Olympic Mazama pocket gopher	Meadows, young forest (North and northeast)	Verts and Carraway 2000

Table 2.1 (continued)

Scientific name	Common name	Habitat association (and quadrant on Peninsula)	Source (for habitat association)
Mustela erminea olympica	Olympic ermine	Throughout (North, central, and southeast)	Hall 1945
Fish			
Novumbra hubbsi	Olympic mudminnow	Low gradient rivers, muddy sediment (South)	McPhail 1967
Opiliones (harvestmen)			
Acuclavella makah		Perennial headwater stream banks, woody debris	Richart and Hedin 2013
Orthoptera (grasshoppers)			
Nisquallia olympica	Olympic grasshopper	Subalpine to alpine, scree sites	Rehn 1952
Lepidoptera (butterflies and moths)			
Hesperia comma hulbirti	Hulbirt's skipper	Subalpine to alpine	Lindsey 1939
Oeneis chryxus valerata	Olympic arctic	Subalpine to alpine (Northeast)	Yake 2005
Coleoptera (beetles)			
Bryelmis rivularis	Riffle beetle	Streams 3–6 m wide, woody debris (West and NW Oregon)	Barr 2011
Cicindela bellissima frechini	Pacific coast tiger beetle	Sand dunes and deflation plains (West)	Leffler 1979
Nebria acuta quileute	Quileute gazelle beetle	River banks at mid-elevation (North)	Kavanaugh 1979
Nebria danmanni	Mann's gazelle beetle	Montane to subalpine (Northeast)	Kavanaugh 1981
Diplopoda (millipedes)			
Leonardesmus injucundus	–	Litter and soil, dense forest (South)	Shelley and Shear 2006
Microlympia echina	–	Riparian forest, alder litter (West)	Shear and Leonard 2003
Tubaphe levii	Olympic peninsula millipede	Lowlands (West)	Causey 1954
Mollusks			
Hemphillia burringtoni	Arionid jumping slug	Dense riparian forest, possibly also in the Cascade Range	Burke 2005

[a] Welch (2008) found no phylogenetic support for a separate clade for *Scapanus townsendii olympicus*

Fig. 2.4 Endemic species richness in 4-km grid cells. This map was developed from Fig. 2.3

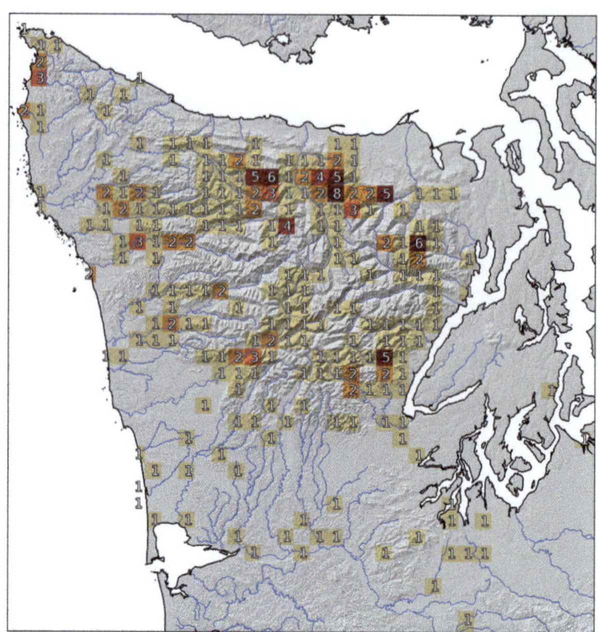

such as on unglaciated mountainsides above the Cordilleran ice sheet (Buckingham et al. 1995).

Although not strictly endemic to the peninsula, seaside juniper (*Juniperus maritima*) exemplifies the association of endemics with northeastern subalpine ecosystems. This species was recently distinguished from *Juniperus scopulorum*, and it has a distribution pattern and genetics that are strongly suggestive of a glacial refugium in northeastern Olympic meadows (Adams et al. 2010). This species is normally associated with rocky outcrops at sea level in northern Puget Sound and Strait of Georgia, but was found to extend to 1700 m above sea level in meadows in the northeastern Olympic Peninsula. Genetic markers show that mid-elevation *Juniperus* on the peninsula are ancestral to more recently dispersed populations in Puget Sound (Adams et al. 2010). Thus, while this species is not strictly endemic to the peninsula, much of its genetic diversity is endemic and suggests unglaciated areas in the eastern peninsula served as glacial refugia.

Several other species are restricted to dry sites on the Olympic Peninsula with the nearest conspecific common in the eastern Cascades and/or Rocky Mountains (e.g., Englemann spruce, whitebark pine, and quaking aspen). Kuramoto and Bliss (1970) suggest that the phlox-bunchgrass community of the northeastern subalpine meadows is a remnant of the early Holocene warm period. However, the high number of endemic taxa of this vegetation type suggests that there was a longer history of persistence underlying these distribution patterns. Consistent with this longer history of dry conditions is the finding of Cactaceae pollen dating to 13,000 years ago near Sequim, indicating that the local prickly pear cactus has existed on the northeastern peninsula since shortly after deglaciation (Petersen et al. 1983).

Nine endemic taxa are associated with stream and shaded sites on the southwestern Olympics. The Olympic mudminnow occurs on the southern peninsula, possibly colonized from a Chehalis River refugium after deglaciation (McPhail 1967), in combination with refugia along the western coast (DeHaan et al. 2013). In contrast, genetics of disjunct populations of mudminnows in eastern Puget Sound (not mapped in Fig. 2.3) suggests these are recent introductions from the southern Olympic Peninsula rather than the result of vicariance by the Puget Lobe of the Cordilleran Ice Sheet (DeHaan et al. 2013). Other shade and stream-associated taxa include little-studied arthropods with strong associations with perennial streams, such as riffle beetles (Barr 2011), harvestmen (Richart and Hedin 2013), and millipedes (Shear and Leonard 2003).

Given the rate of discovery, there are likely many more endemic lineages that suggest persistence through glacial cycles. Arthropod diversity, for example, is little described (e.g., Winchester and Ring 1999). Two frog species also show endemic clades. The Cascade frog (*Rana cascadae*) of the Olympic Peninsula, though not an endemic species to the peninsula, contains a distinct mtDNA lineage that suggests isolation through glacial periods (Monsen and Blouin 2003). The tailed frog (*Ascaphus truei*) also contains a distinct Olympic clade (Nielson et al. 2001), and is associated with similar habitats as the endemic Olympic torrent salamander (*Rhyacotriton olympicus*; Adams and Bury 2002). Similarly, Van Dyke's salamander (*Plethodon vandykei*) and Cope's giant salamander (*Dicamptodon copei*) feature co-distributed disjunctions in western Washington with unique Olympic clades, though details of the phylogeographic patterns suggest different dispersal histories of each taxon (Steele and Storfer 2007). Though discoveries continue, the current list of endemic taxa that both have a low capacity for dispersal (vagility) and that are associated with the moist lowland habitat suggests that this habitat type, namely the perennial streams and forest understory habitat, has existed through at least one glacial cycle on the peninsula, or at least nearby in southwest Washington.

Another set of peninsula taxa are disjunct by long distances (hundreds of kilometers) from the remainder of their population, or for endemics, to their sister taxon, thus also supporting the existence of a Pleistocene refugium. This list of species is long and difficult to compile, but generally falls into two groups. One group is a pattern of distribution on the peninsula and in the boreal and tundra biomes of the Yukon and Alaska. The ranges of these species likely split during the Pleistocene glaciations, with the Olympic Peninsula constituting an important southern refugium. An example of this pattern is the long-stalked whitlow grass. The second pattern is a distribution on the peninsula that is disjunct from an interior Rocky Mountain distribution and absent from the Cascade Range. Examples of this pattern are the least-bladdery milkvetch (*Astragalus microcystis*) and western sweetvetch (*Hedysarum occidentale*; Houston et al. 1994). Taxa that have low vagility and strong habitat associations with wet forest habitat often have a sister species occurring in similar wet forests of northern Idaho. This disjunct pattern is common between the Olympic

Peninsula and North Idaho, being found for riffle beetles, salamanders, harvestmen, and several other taxa (Table 2.1 and reviewed in Gavin 2009).

The isolation of the Olympic Peninsula has resulted in several species that have not naturally colonized the peninsula from the Cascade Range. Notable mammals and birds in this group include grizzly bear, wolverine, red fox (introduced in the twentieth century), coyote (now present), lynx, water vole, golden-mantled ground squirrel, northern bog lemming, porcupine (now entering the peninsula), pika, mountain sheep, mountain goat (introduced in the twentieth century), and the white-tailed ptarmigan. Absent trees include noble fir, ponderosa pine, subalpine larch, western larch, and western juniper (Houston et al. 1994). Other species may have become extinct in the Pleistocene, such as a trechine beetle discovered in the beach cliff at Kalaloch (Cong 1997).

In contrast to the insularity of the Olympic Peninsula, some biogeographic patterns reflect its limited connection to the archipelago of land bridge islands to the north (connected during lower sea levels). Buckingham et al. (1995) highlight several plant species that are endemic to the land bridge islands with their southern limit on the Olympic Peninsula, and thus suggest a northward expansion from an Olympic refugium. Six plant species are endemic to only the Olympic Peninsula and Vancouver Island, and dozens more spread further north along the coast. For well-dispersed species, north–south migration along the archipelago may have tracked climate change fairly closely. Marine taxa may have changed their distributions most readily. For example, a female walrus skeleton dating to ca. 70,000 years ago was found at Qualicum Beach on Vancouver Island, suggesting that during cold stadial periods of the late Pleistocene, the region was similar to the modern Bering Sea (Harington 2008).

A biogeographic pattern dubbed the "peninsula effect" is that species diversity is expected to decline from the base to the tip of a peninsula. For example, George Gaylord Simpson (1964) mapped North American mammal richness and found that major peninsulas contained markedly few species than areas of equal size in the interior. Though Simpson's coarse-resolution data could not resolve the Olympic Peninsula, this pattern would certainly be upheld on the peninsula at a finer resolution. A similar mapping of bird species richness by Cook (1969) invoked the peninsula effect that results from immigration and extinction dynamics (but like Simpson, Cook did not resolve patterns on the Olympic Peninsula). More recent work has shown that the peninsula effect rarely exists as a result of land geometry, especially for vagile species such as birds and winged insects, and richness gradients are mostly explained by habitat as controlled by physiographic factors (Wiggins 1999). This same explanation appears to largely hold for the set of Olympic endemic taxa (Fig. 2.4). The history of glaciation and continuation with coastal mountains to the north results in continuity of many species distributions off the tip of the Olympic Peninsula to the north.

In summary, the biogeographic distributions of several plant species and communities suggest the existence of glacial refugia in the north and northeastern Olympic Mountains (Houston et al. 1994). We note that both the alpine northeastern mountains and southwestern peninsula contained substantial unglaciated areas during the glacial periods. However, many endemic taxa are subspecies or closely related to taxa off of the peninsula and thus are likely "neoendemics" that differentiated since deglaciation. Other endemics that are far from their sister taxa both geographically and evolutionarily date to deeper evolutionary splits and may be considered "paleoendemics." The potential that such endemics remained in situ through the major climate changes of the Late Quaternary is intriguing but strong support for this requires both phylogeographic studies and a better understanding of past climate and vegetation at fine (microclimatic) resolution (Hampe and Jump 2011).

Part II
Postglacial Paleoclimate and Environmental Change on the Olympic Peninsula, Washington

Chapter 3
Postglacial Climate on the Olympic Peninsula

Abstract Paleoecological studies reveal how biota responded to past climate changes, some of which were rapid and others of which were gradual. Hence, it is essential to understand the climate changes that drive the biotic response. This chapter begins with a concise review of the global-to-local causes of climate change since the Last Glacial Maximum. It then presents a review of paleoclimatic change on the Olympic Peninsula during and following deglaciation. Paleoclimatologists use a variety of proxies and attribute the fluctuations in proxies to certain climate parameters (e.g., summer drought). Through careful analysis of the literature, this chapter presents a "best-evidence" synthesis of climate proxies taken from the broader Pacific Northwest region but with clear implications for understanding changing environments on the peninsula. Changes in vegetation and fire occurrence are purposely avoided in this chapter because they are responses to climate change and may lag behind actual climate changes by more than 100 years. In contrast, other proxies of climate, such as the temperature tolerances of beetles or midges or the isotopic composition of carbonates, should respond with much shorter lag times and are normally more straightforward in their climatic interpretation compared to vegetation proxies. A particularly useful climate proxy in the Pacific Northwest is the record of glacial advances and retreat. This chapter is presented in chronological order with a focus on regional coherence among various climate proxies, timing of events, and which seasonal temperature or moisture variables underlay the observed climate proxies.

3.1 Forcing Factors in the Climate System

The three major controls of Earth's climate—incoming solar radiation (insolation), ice extent, and greenhouse gases—have changed dramatically from the glacial maximum to the present (Fig. 3.1). Earth's orbit varies slowly on cycles lasting from 20,000 to 100,000 years. These Milankovitch cycles involve the shape of the orbit (eccentricity), the tilt of Earth's axis (obliquity), and the seasonal precession of the perihelion, or closest Earth–Sun distance. All of these combine to affect the latitudinal distribution and seasonal cycle of incoming solar radiation (insolation). These cycles are the pacemaker of glaciations, as the onset of ice sheet growth occurs

© Springer International Publishing Switzerland 2015

D. G. Gavin, L. B. Brubaker, *Late Pleistocene and Holocene Environmental Change on the Olympic Peninsula, Washington,* Ecological Studies 222,

DOI 10.1007/978-3-319-11014-1_3

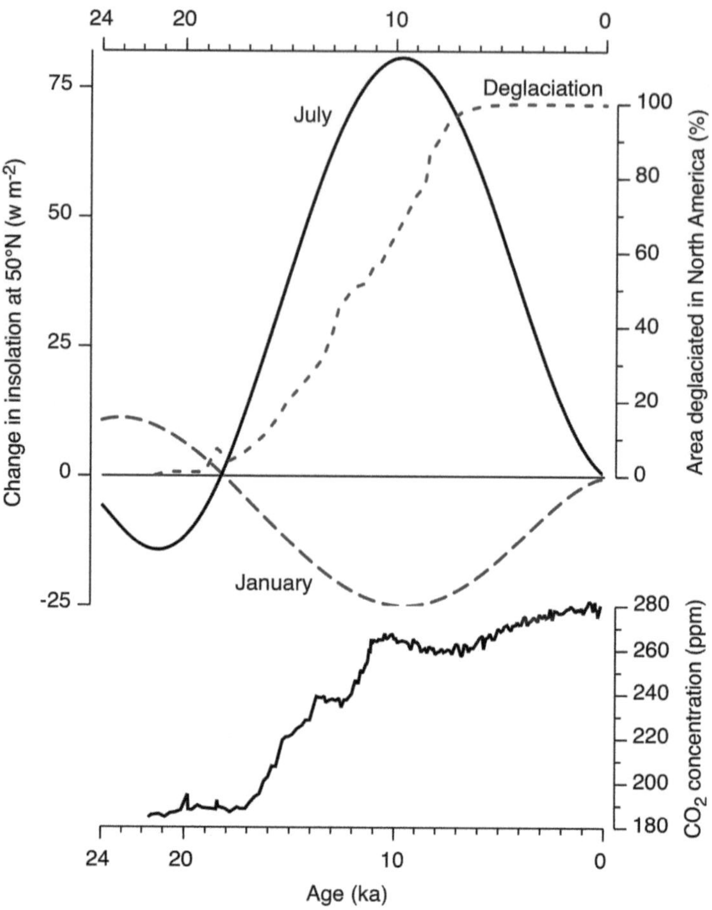

Fig. 3.1 Major forcing factors of climate change for the past 21,000 years. S: Summer (JJA: June, July, August) and winter (DJF: December, January, February) insolation anomalies for the northern hemisphere; CO_2 concentrations and land ice (percent of maximum cover) are also shown. (Modified from Kutzbach et al. 1998)

when northern hemisphere summer insolation is low and deglaciation accelerates when summer insolation is high. The extent of glacial ice sheets provides a positive feedback in the climate system. Extensive ice sheets during the Last Glacial Maximum (LGM; 24–19 ka or thousands of years before present[1]) reflected short-wave solar radiation back to space, altering the global balance of incoming and outgoing radiation, as well as modifying the locations of atmospheric high- and low-pressure systems, the location of jet streams, and routes of moisture to continental interiors. Greenhouse gases, including carbon dioxide and methane, were lower during the LGM, thus increasing the amount of outgoing long-wave radiation escaping from

---

[1] All radiocarbon ages are calibrated to calendar years using a recent calibration curve (Reimer et al. 2009).

Earth's surface and atmosphere into space. The large-scale controls on Earth's climate were successfully simulated in the Cooperative Holocene Mapping Project (COHMAP), which provided the first strong link between empirical data and climate models on the climates since the LGM (COHMAP Members 1988; Kutzbach et al. 1998).

In addition to these well-known forcing agents of climate change, other components of the climate system add shorter-term variability, resulting in a complex history that is marked by variability occurring over a wide range of timescales. Solar forcing due to variation in solar intensity, although of low magnitude, may become amplified by affecting upper-atmospheric ozone concentrations. Volcanic activity may shield insolation for several years. Changes in large-scale features of ocean and atmospheric circulation may experience sudden shifts, especially during deglaciation, producing abrupt climate changes, some of which may persist for decades to millennia (Gavin et al. 2011). During the glacial period, these fluctuations were pronounced and are referred to as stadial (cold) and interstadial (warm) events.

3.2   The Late Pleistocene and the Last Glacial Maximum: >60–19 ka

The late Pleistocene, defined as the period up through the LGM to the Holocene (11.7 ka to present), was a period of alternating cold glacial advances and warmer periods (Fig. 3.2). Large valley glaciers on the western peninsula repeatedly advanced and formed extensive moraines, one of which created Lake Quinault (Fig. 3.3). Radiocarbon dates on these moraines indicate maximum glacier extents older than 50 ka (Lyman Rapids advances), ca. 38 ka (Hoh Oxbow I advance), and ca. 32 ka (Hoh Oxbow 2 advance, which was the greatest ice extent of the western Olympics during the late Pleistocene) and ca. 22 ka (Twin Creeks 1 advance; Thackray 2001; Thackray 2008). These events correlate broadly with alpine glaciers in the Skagit Valley (Riedel et al. 2010) and on Mount Rainer (i.e., the Evans Creek Stade; Booth et al. 2003). The climate of the period between glaciations was not much colder than present. Fossil beetles can be used to reconstruct past temperatures based on the tolerances of the species in an assemblage. A fossil beetle assemblage from a beach cliff at Kalaloch showed that the latter part of the Olympia Interglaciation, 48–40 ka, was only about 1 °C colder during July compared to present (Cong and Ashworth 1996).

Glacial activity increased the supply of sediment to the rivers, which created broad floodplains and carved into hillslopes of the western Olympic Mountains. Rivers on the western Olympics have now incised through these floodplains and into bedrock, leaving them as strath terraces (Pazzaglia and Brandon 2001). Despite these glacial advances on the western Olympics, there is little evidence of advances of Cordilleran ice into Puget Sound. Evidence for such advances would be difficult to find due to the subsequent Vashon advance (Booth et al. 2003).

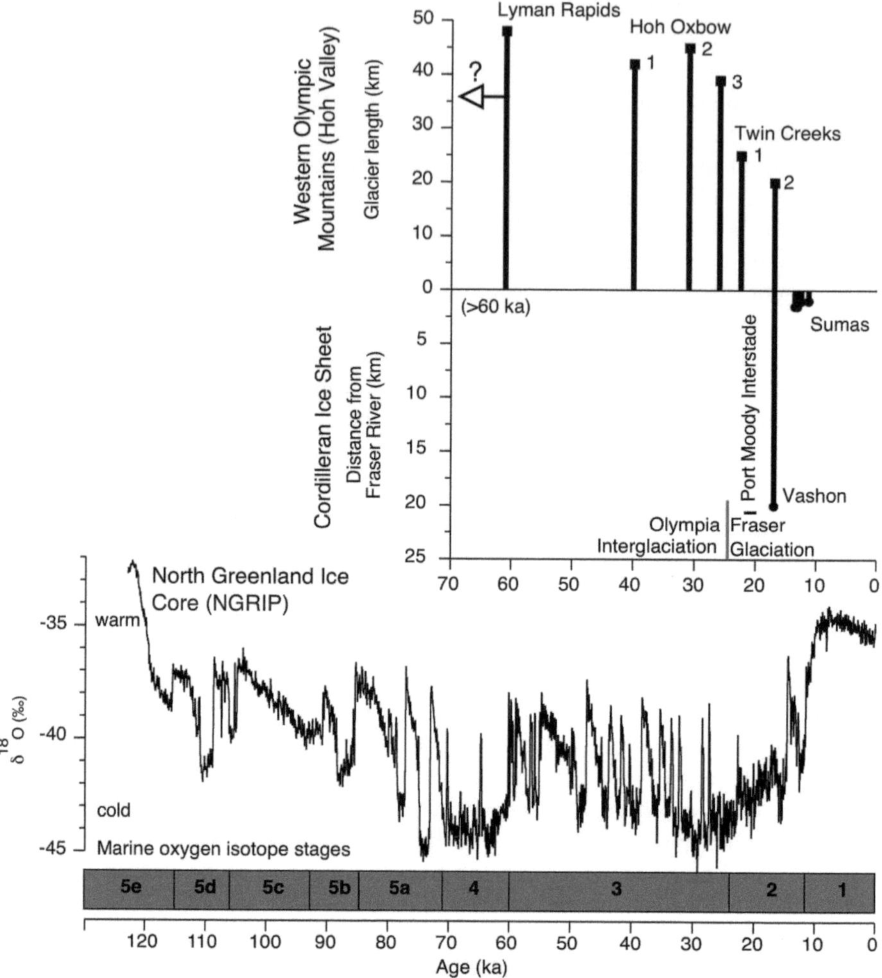

Fig. 3.2 *Top*: Extents of late Pleistocene glacial advances in the Hoh Valley of the western Olympic Mountains and of the Cordilleran Ice Sheet in the Puget Sound. Dates are from Booth et al. (2003) and Thackray (2008). *Bottom*: The Greenland oxygen isotope record (North Greenland Ice Core Project members 2004) indicating temperature in Greenland at 50-year intervals to the last interglacial (marine oxygen isotope stage 5e). Note that the dates of the Hoh glacial advances generally date to stadial (cold) periods during stage 3 and that the very extensive Lyman Rapids glacial advance falls during the cold stage 4

During the LGM, the seasonal pattern of insolation was similar to the present, but extensive ice sheets were located to the north and atmospheric CO_2 was as low as 180 ppm (65 % of preindustrial levels). The high albedo and high elevation of the Cordilleran and Laurentide ice sheets produced very cold air. This cold air resulted in a strong high-pressure system that produced a clockwise (anticyclonic) air flow, which was especially strong during winter. The anticyclonic winds spun off the ice

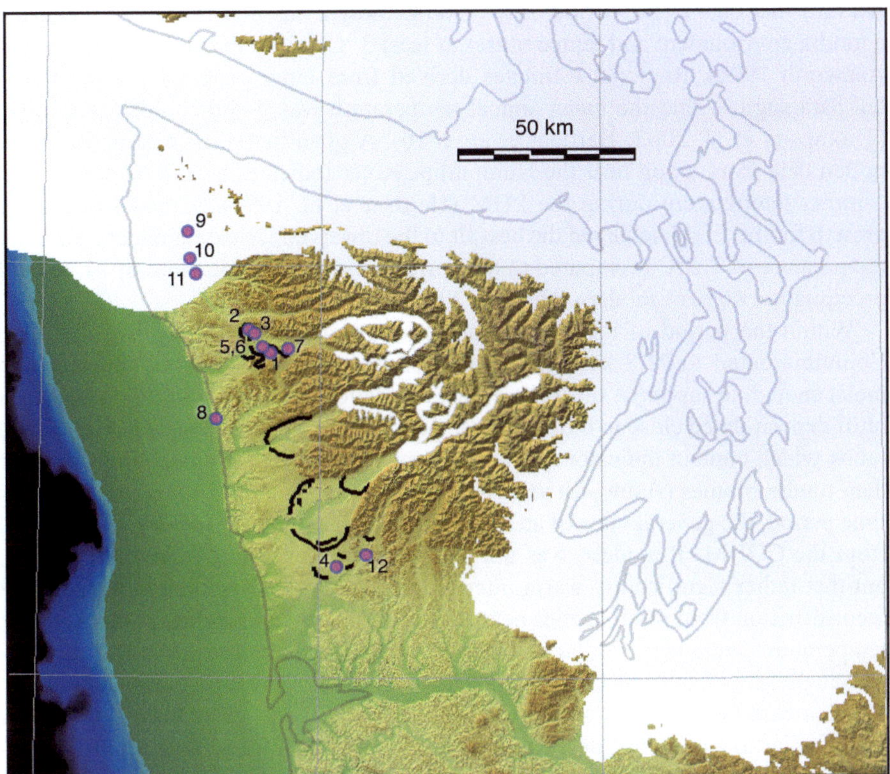

Fig. 3.3 The maximum extent of the Vashon Stade advance into Puget Sound and the Strait of Juan de Fuca, paleogeography based on a 120-m decrease in sea level, and modern coastlines shown by a *gray line* (though this is an overestimate as it does not consider isostatic depression, see James et al. 2009). Alpine glaciers show extent of the Twin Creeks 2 Stade (Thackray 2001), which was likely coeval with the Vashon Stade of the Cordilleran Ice Sheet (17 ka). *Black lines* show moraines dating to the maximum late Pleistocene extent of the valley glaciers in the Hoh, Queets, Quinault, and Humptulips rivers, dating to ca. 23 ka. Glacier extents are from Porter and Swanson (1998) and Thackray (2001); numerous alpine glaciers are not mapped. *Numbered sites* are lakes and bogs studied by Cal Heusser and discussed in Chap. 4: *1–3* Hoh River Valley sites 1–3 (Heusser 1964); *2* Bogachiel Bog (was also studied by Hansen 1941 and Heusser 1978); *4* Humptulips Bog (Heusser 1964; Heusser et al. 1999); *5–7* Hoh bogs 1–3 (Heusser 1974); *8* Kalaloch sea cliff (Heusser 1972); *9–11* Wessler Bog, Wentworth Lake, and Soleduck Bog, respectively (Heusser 1973)

sheets producing an east-to-west wind south of the glacial limits in Washington State; the Loess hills of eastern Washington date to the Pleistocene when these winds were strong. The westerly flow of the winter jet stream was thus diverted to the south, resulting in less annual precipitation than occurs today (Kutzbach et al. 1998).

The climate of the LGM was colder than present, though few quantitative climate reconstructions date to this time period. Beetle assemblages from the Kalaloch sea cliff that date to the period of the Hoh Oxbow 2 advance up to 19 ka indicate a tundra environment and temperatures at least 3 °C colder than present (Cong and Ashworth 1996). Regional estimates derived from large geographic sets of pollen data suggest that the mean annual temperature was 5–8 °C colder than today (Thompson et al. 2003; Bartlein et al. 2010). A quantitative reconstruction from pollen data from a bog near the Humptulips River indicates a ca. 5 °C decrease in summer temperature during the LGM (Heusser et al. 1999). A model of glacier growth for the LGM achieved the best fit to the observed ice extent under a summer temperature of 7.2 °C (compared to the current sea-level temperature of 14 °C) and precipitation 40 % of modern (Hellwig 2010).

Within the period of the LGM, a short interstadial event in southwest British Columbia dated to 21.8 ka, dubbed the Port Moody interstadial, was warm and moist enough to support a subalpine forest vegetation (Lian et al. 2001). In Seattle, a bluff deposit dated close to the same time period (19.8–18 ka) revealed a rich beetle fauna which contain indicator species that are associated with forest biomes rather than tundra biomes (Ashworth and Nelson 2014). Both studies found, that for this time period, the general view of an LGM climate that was cold and dry, as predicted from the COHMAP models, was not generally correct through the entire period, but that rather significantly warm interstadials occurred. Quantitative temperature reconstruction from fossil beetles near Seattle dating to 18 ka indicates that summer temperatures were very similar to today, but winter temperatures were 7–8 °C colder than present (Ashworth et al. 2000; Ashworth 2003; Ashworth and Nelson 2014). This increased continental climate is consistent with a vegetation of subalpine parkland similar to a modern Engelmann spruce-subalpine fir forest (Lian et al. 2001).

3.3 The Late-Glacial: 19–11.6 ka

The period of deglaciation at the end of the Pleistocene, between 19 and 11.6 ka, involved fluctuations on timescales of centuries, decades, and years, as well as abrupt long-term changes. During this period, summer insolation and atmospheric CO_2 was increasing (Fig. 3.1). The period is marked by a complex series of climate changes when the location of the jet stream was adjusting as the ice sheet was waning, changes that also resulted in large outburst floods that affected ocean circulation. As the ice sheets retreated north, the glacial anticyclone weakened and the westerly jet stream moved northward, causing increased onshore flow and increased precipitation in the Pacific Northwest. An associated large increase in moisture at 17 ka caused glacial advances throughout the Pacific Northwest. This resulted in the maximum extent of Cordilleran ice during the Vashon advance into Puget Sound (Fig. 3.3). This advance correlated in time to the last major advance of valley glaciers on the Olympic Peninsula (the Twin Creeks II advance of Thackray 2001). The Vashon lobe had completely retreated by 16 ka. This rapid retreat was facilitated by calving into a proglacial lake dammed south of the ice front (Porter and Swanson 1998). Most subalpine cirque lakes in the Pacific Northwest date to 14.5 ka, indicating that alpine glaciers did not retreat to their near-modern extent until 1500 years after the Vashon retreat. The loss of the alpine glaciers occurred very close in time to the onset of the Bølling–Allerød warm interstadial event (14.7–12.9 ka) that is well known in Europe and is documented from marine paleoclimate records in the Pacific Northwest (Kienast and McKay 2001; Barron et al. 2003).

The next major climatic event was the Younger Dryas stadial between 12.9 and 11.6 ka. This cold period, which is a distinct event in the North Atlantic, has been identified to various degrees in the Pacific Northwest. Mathewes (1993) and Mathewes et al. (1993) synthesized palynological data from lake sediment and benthic foraminifera from ocean cores along the British Columbia coast, revealing a cold event synchronous with the Younger Dryas stadial that is well known in the North Atlantic. The event was thus shown to be hemispheric in scale and involve both the ocean and atmosphere. More recent well-dated climate records from a speleothem (stalagmite) in Oregon Caves (Vacco et al. 2005) and the organic content of lake sediments on the Olympic Peninsula (Gavin et al. 2013) are synchronous with the Younger Dryas. A reconstruction of sea-surface temperatures (SST) indicates a nearly 4 °C cooling during the Younger Dryas relative to today, while a July temperature reconstruction from British Columbia based on fossil midges (chironomids) reveals a more muted cooling of ca. 2 °C (Fig. 3.4).

A re-advance of Cordilleran ice in southwest British Columbia during the late-glacial, termed the Sumas Stade, has generated much discussion. A series of moraines in the Fraser River Lowland date to 13.7–13.2 ka (Clague et al. 1997; Kovanen and Easterbrook 2002), which agrees with a moraine age of 13.5 ka in a nearby fjord (Menounos et al. 2009) and a glacier advance at Mt. Rainier dated to before 13.2 ka (Heine 1998). Another moraine at Howe Sound north of Vancouver dates to the earliest part of the Younger Dryas at 12.8 ka (Friele and Clague 2002; Menounos et al. 2009). Kovanen and Easterbrook (2002) also report two minor re-advances that date to within the Younger Dryas Chronozone. However, a detailed pollen record from near the Fraser Lowland indicates that the Younger Dryas climate gradually warmed and then cooled, but the cooling was too late to be consistent with the latter Sumas advances (Pellatt et al. 2002). A further discussion provides some doubt that the chronologies of some of the Younger Dryas glacial re-advances are correct (Easterbrook 2004; Pellatt et al. 2004). Several moraines in the southern Coast Mountains of British Columbia and Mount Baker have poorly constrained ages that may date to the Younger Dryas (Osborn et al. 2012). Regardless of the exact chronology, several glacial advances occurred in the southern Coast Mountains between 13.7 and 12.8 ka.

3.4 The Early Holocene: 11.6–6 ka

Increased summer insolation during the Holocene led to an intensified Pacific subtropical high-pressure system that created warm, stable, dry air to its east (i.e., the Pacific Northwest). At the same time, decreased winter insolation may have

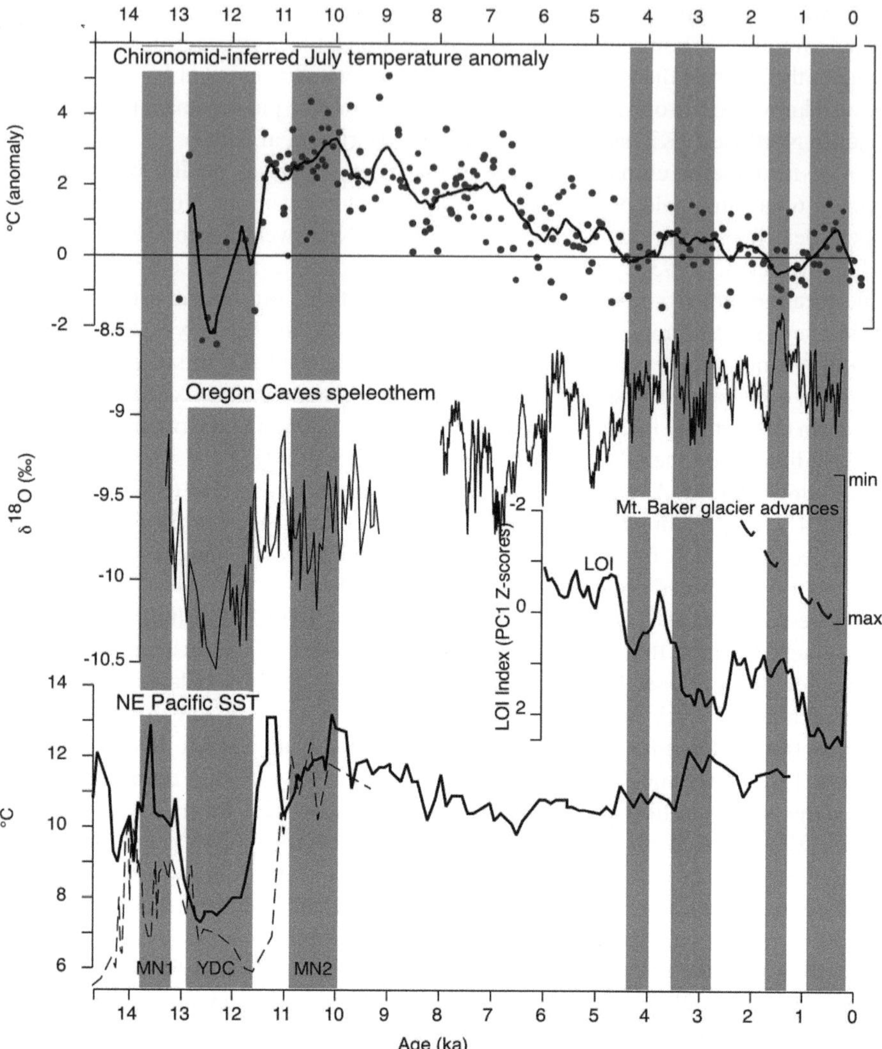

Fig. 3.4 Paleoclimate indicators from the Pacific Northwest, adapted from Gavin et al. (2013). The chironomid-inferred July temperature anomaly is based on four lakes in southern British Columbia (Palmer et al. 2002; Rosenberg et al. 2004; Chase et al. 2008). *Points* indicate the temperature anomaly from a 0–2-ka base period, and the *smooth line* shows a locally weighted (loess) smoothing in a 600-year window and its standard error (*gray lines*). The loss-on-ignition (LOI) index is from glacial-fed lakes in southwest British Columbia that are sensitive to times of glacier advances (Menounos et al. 2008). Times and relative magnitude of five glacial advances on Mount Baker, Washington, are overlaid on the LOI index (Osborn et al. 2012). The speleothem oxygen isotope (δ18O) record is from Oregon Caves National Monument (Vacco et al. 2005; Ersek et al. 2012). The Northeast Pacific sea-surface temperature (SST) is inferred from alkenone records from a core near northern California (Barron et al. 2003), which broadly agrees with a lower-resolution record closer to the Olympic Peninsula (*dashed line*; Kienast and McKay 2001). *Vertical shading* indicates times of glacial advances described by Heine (1998) for the McNeeley (MN) 1 and 2 and by Menounos et al. (2009) for the advances from 4.2 ka to present. The Younger Dryas Chronozone (YDC), possibly not a period of glacier advance in the region, is also shown

resulted in winter temperatures colder than today and thus an overall more continental climate with greater seasonal cycle of temperature. This period has been broadly termed the "hypsithermal," the "xerothermic," or the "Holocene climatic optimum." Fossil chironomid remains recovered from lake sediments in southern British Columbia suggest that early Holocene summer temperatures were 4 °C warmer than present, with the warmest temperature broadly between 11.5 and 10 ka (Rosenberg et al. 2004; Chase et al. 2008; Fig. 3.4). In contrast, SST off of northern California were only 1–2 °C warmer than present, possibly due to the moderating influence of upwelling during the early Holocene resulting from a moderately strong California Current (Barron et al. 2003). The Oregon Caves oxygen isotope record, in which less negative values indicate warmer temperature, does not track early Holocene summer warmth (Fig. 3.4). Rather, this proxy is sensitive to temperatures during the rainy season, and in particular the early winter months (Ersek et al. 2012). Thus, the lack of early Holocene warming indicated in the Oregon Caves record is consistent with increased seasonality during early Holocene times.

The early Holocene summers were likely not uniformly warm and dry, but rather marked by distinct century-scale periods of increased moisture. The remnants of the waning ice sheet to the north may have still influenced the jet stream across western North America (Gavin et al. 2011). For example, dated lake sediments from Mount Rainier suggest a glacial advance between 11 and 10 ka (McNeeley 2 advance of Heine 1998), which is synchronous with a cooling of SST and cooler temperatures indicated in southern Oregon (Fig. 3.4). Gavin et al. (2013) have found forest compositional changes on the Olympic Peninsula that support increased moisture between 11 and 10 ka. However, this interpretation does not entirely agree with the chironomid-based temperature reconstruction, though dating errors may prevent detection of short-term minor temperature change. Because similar glacial advances have not been found elsewhere in western Washington, the validity of the McNeeley 2 advance has been questioned (Menounos et al. 2009).

3.5   The Middle-to-Late Holocene: 6 ka to Present

A millennial-scale trend of decreasing summer insolation resulted in a weakening of the Pacific subtropical high and the onset of cooler summer temperatures through the middle Holocene. This latter part of the Holocene is termed the "neoglacial" because many alpine glaciers began advancing downslope about 4000 years ago. Many of these glacial advances, including an event at 3.3–2.9 ka, were synchronous across the West (Fig. 3.4). Of all the glacial advances during the Holocene, almost without exception the largest was a series of Little Ice Age glacier advances from 1350 to 1850 AD (Osborn et al. 2012). These fluctuations are also quite synchronous with centennial-scale variations in the Oregon Caves oxygen isotope record, which is most sensitive to winter-season temperature (Ersek et al. 2012). The chironomid-inferred temperature curve shows little variability through this period. However, some vegetation and fire-regime changes were broadly synchronous across the Pacific Northwest, and will be summarized in the following chapters.

The climate changes in western Washington summarized in this chapter were of a scale large enough to affect major components of ecosystems. Of these, important climate variables, such as the spring snowpack and the degree of summer drought, changed by large enough degree to affect the fire regime and the species composition of the forests. In some cases, such as the warm-up into the Holocene, there was nearly a complete turnover of species at a site. In other cases, such as the change from early-to-late Holocene climate, the disturbance regime mediated major changes in the balance of forest species. The next chapter addresses these ecosystem responses to climate in detail.

Chapter 4
Late Quaternary Vegetation and Fire History of the Olympic Peninsula

Abstract The climate and vegetation history of the Olympic Peninsula has been complex over the Late Quaternary. Reconstructing such histories and understanding how they are linked mechanistically is a major goal of Earth system science. Advances are made when there is intensive development of paleo-records and mechanistic models. The Olympic Peninsula has the potential to be such a location, as the area has received attention from the pioneering years of pollen analysis up to the present. First, this chapter reviews the paleoecological studies of climate and vegetation on the Peninsula, including studies that extend back to before the Last Glacial Maximum. Second, a presentation of five detailed pollen records that span across the climatic gradient of the peninsula shows how regional gradients of vegetation have evolved through time. Third, a regional synthesis of vegetation change, including western Washington and southwest British Columbia, is presented through mapped pollen data. This taxon-level analysis provides a view into climatic gradients of the past and suggests modern analogs for periods, such as the late-glacial, that have been difficult to interpret from single sites.

> Climate changes during the late-Quaternary have been complex, and the details of the vegetational response to climate change are also complex.
> Thompson Webb III,
> For every complex problem there is a simple solution. And it is always wrong.
> H.L. Mencken

4.1 Pioneering Studies

The vegetation history of the Olympic Peninsula was first described by pioneering paleoecologists who worked when pollen analysis from lake and bog sediment was a new approach. Henry P. Hansen developed dozens of pollen profiles in the Pacific Northwest while a PhD student at the University of Washington in the 1930s and as faculty at Oregon State University in the 1940s. His work preceded radiocarbon dating, which made it difficult to interpret many aspects of his pollen records. For example, he described a coarse pollen profile from a low-elevation bog near the Bogachiel River (Hansen 1941) that was dominated by mountain hemlock and pine. Radiocarbon dating would have likely indicated he had reached late-glacial

© Springer International Publishing Switzerland 2015
D. G. Gavin, L. B. Brubaker, *Late Pleistocene and Holocene Environmental Change on the Olympic Peninsula, Washington,* Ecological Studies 222,
DOI 10.1007/978-3-319-11014-1_4

sediments, but since there was no chronology to make this interpretation he speculated that pollen had been transported downhill in streams to his bog site.

In Hansen's (1947) monograph *Postglacial Forest Succession, Climate, and Chronology in the Pacific Northwest,* he summarized more than 70 pollen profiles and suggested that the "postglacial climatic trends have been much the same over the entire region, although varying in degree." Edward Deevey, in a review of Hansen's book, was more pessimistic about the claim that such a coherent trend existed in Hansen's data, and instead asserted that the data indicate complex successional patterns strongly influenced by fire and other disturbances (Deevey 1948). Nevertheless, the sequence described by Hansen remains largely valid: cool and moist late-glacial, followed by increased warming and drying until 8000 years ago, followed by maximum warmth, and cooling during the past 4000 years.

As the first report of its kind, Hansen (1947) provides a comprehensive review of the geography, vegetation and lake and bog origins in the Pacific Northwest, as well as pollen morphology, coring methods, and the pollen stratigraphy of several dozen sites. In the Puget Lowland, he describes the replacement of lodgepole pine by Douglas-fir as climate "ameliorated" in the late-glacial and then the increase in western hemlock after the deposition of the Mazama tephra (the Crater Lake eruption that is now aged to 7627 ± 100 years ago). Hansen acknowledged that this sequence could be due to a long protracted ecological succession on the deglaciated landscape, as this sequence of species follows an order of increasing shade tolerance and increasing requirements for organic and moist soils. However, Hansen also noted some fluctuations in western hemlock that were assumed to be synchronous and are not expected due to succession, therefore favoring a climate-change explanation. Furthermore, he equated the early Holocene vegetation in the Puget Sound to that of the Willamette Valley of Oregon today. In contrast, along the coastal strip, Hansen found very little evidence for vegetation change during the Holocene and suggested that the hyper-maritime climate along the coast is buffered from regional climate change. These sites were from the Oregon coast where the relative stasis of Holocene vegetation change has been confirmed by subsequent studies (Long and Whitlock 2002). Hansen conducted no further studies from the western Olympic Peninsula.

4.2   The Late Pleistocene

Following Hansen's pioneering work, Calvin J. Heusser conducted two decades of research on the vegetation and glacial history of the western Olympics. In 1960, he published the monograph *Late Pleistocene Environments of North Pacific North America,* which summarized the state of knowledge of tree biogeography and climate history of that area (Heusser 1960). Heusser focused on dozens of mostly undated pollen diagrams from southeastern Alaska but also included preliminary results from Washington and Oregon.

Heusser's first major study in Washington was the pollen analysis of three peat bog sections near the Hoh River and one longer section near the Humptulips River

(Heusser 1964). The three pollen records from the Hoh, two of which had basal dates of ca. 15,000 ^{14}C yr BP (before present), showed the now well-known progression of pollen types of grass and pine and mountain hemlock transitioning to alder and eventually to western hemlock and Sitka spruce. The longer Humptulips record, nearly 6.3 m of fibrous and sphagnum peat and lake sediment, extended back to before 50 ka. Before the Holocene, the record was marked by fluctuating levels of pine (likely shore pine), with varying levels of Sitka spruce, grass, and sedges. Based on a data set of modern pollen assemblages northward along the Pacific coast, Heusser (1964) assigned pollen analogs to fossil samples to reconstruct the history of temperature and precipitation. These curves indicated a 10 °F decrease in mean July temperature, a decrease in precipitation, and fluctuations that crudely match the hypothesized sequence of glacial advances during the late Pleistocene.

Heusser (1964) remarked that the Olympic Peninsula vegetation record was very dynamic relative to other locations on the Pacific coast. He also noted that the Humptulips pollen diagram indicated the clear presence of trees through the peak of the Last Glacial Maximum (LGM; the "Late Wisconsin"), though sampling was coarse and dating control was marginal by today's standards. In particular, the LGM vegetation was interpreted as open parkland of mountain hemlock with some subalpine fir, Sitka spruce, shore pine, and a rich herbaceous community. The earlier "Middle Wisconsin" had fewer tree species and a valley glacier within 2 km of the site. Heusser's (1964) results support the assertion that the southwestern Olympic Peninsula functioned as a refugium for tree species through the glacial periods. He also speculated that such refugia did not likely extend northward into the purported small unglaciated areas north of Washington, though the possibility of more northern refugia along the Pacific coast remains an area of study (Shafer et al. 2010; Elias 2013).

Heusser later developed a longer and more detailed late Pleistocene pollen record from a beach cliff exposure almost 30 m in height near Kalaloch, Washington (Heusser 1972). This pollen record is marked by large fluctuations between tree and herb pollen types, suggesting frequent changes in forest density related to temperature change and submillennial-scale fluctuations not yet inferred from the glacial chronology. The upper part of the beach cliff reveals the transition from a mountain hemlock parkland prior to the LGM to a mostly treeless environment during the LGM. However, a brief peak in western hemlock pollen coincided with the Port Moody Interstadial period (ca. 21 ka) in southwest British Columbia, interrupting the period of low tree cover (Lian et al. 2001). Earlier in the Pleistocene, during the Olympia Interglaciation, western hemlock and Sitka spruce pollen were abundant, suggesting a nonglacial interval with an environment similar to modern montane forests. The Kalaloch chronology was reinterpreted by Thackray (2001) to extend back to the last interglacial period, between 105 and 125 ka, rather than the 70 ka estimated by Heusser. Regardless of the ages of the stadial (glacial) events in this section, the pollen data indicate that during these cold periods some trees persisted in the area, albeit at low density, and that a high level of plant diversity persisted in this coastal refugium within tens of kilometers of extensive glaciation (Heusser 1972). This history also provides important support for the hypothesis that low-vagility endemic arthopods and amphibians of the western Olympic Peninsula persisted in refugia not far from their current location (e.g., Richart and Hedin 2013, Chap. 3).

Additional pollen records obtained near the Hoh River, which is located south of the Juan de Fuca lobe but within mid-Wisconsin recessional moraines, provide a view of the Pleistocene refugium much closer to the glacial limits than seen at Kalaloch (Heusser 1974). One of the three bogs, Hoh Bog, contained a sequence dated to ca. 21 ka, which revealed that a tundra vegetation with only ca. 10 % lodge-pole pine pollen dominated until the Holocene increase in warmth. Heusser (1974) noted that all of the pollen types in the diverse tundra period in the Hoh Bog have modern representatives on the peninsula, with the exception of *Selaginella selaginoides,* which is now found on northern Vancouver Island.

The combination of the Hoh peat core and the Kalaloch section provides a continuous pollen record spanning from before 50 ka to present (Heusser 1977). The interpretation of this sequence was partially confirmed by a peat profile near the Bogachiel River (Heusser 1978). The peat in this section is bracketed by radiocarbon dates between 30 and 20 ka, and the pollen record indicates mountain hemlock parkland transitioning into a tundra-like vegetation during the LGM. However, the rapid transition from the LGM to one typical of the early Holocene suggests that much of the late-glacial (16–12 ka) is missing from this record.

The combined Hoh and Kalaloch pollen record also provided Heusser an opportunity to use pollen data with multivariate statistics to reconstruct the Late Quaternary temperature record. Heusser et al. (1980) compared the long pollen record with a network of modern pollen assemblages from northwest California to south-central Alaska in order to reconstruct mean July temperature and annual precipitation at the fossil pollen sites. This method assumes that modern vegetation is in equilibrium with climate (i.e., there are no migration lags or other factors affecting tree distribution other than climate), which is still debated, and that fossil and modern pollen assemblages are similar enough to represent similar vegetation and climate. By extracting four principal components from the modern pollen data (which are most related to pine, hemlock, and grass pollen types), transfer functions were constructed relating these new variables to current climate, though no measures of uncertainty were presented. When applied to the fossil pollen record, the transfer functions show that the middle Holocene was slightly (1.5 °C) cooler than present, the early Holocene was similar to present, and the LGM was 4 °C cooler than present. These results considerably underestimate the actual variability currently reconstructed from pollen data (Bartlein et al. 2010) for the region. Bartlein's recent precipitation reconstruction shows a dry early Holocene consistent with evidence of fire and Douglas-fir, and a dry LGM (ca. 70 % of modern precipitation). In addition, both temperature and precipitation are much more variable prior to the LGM than during the Holocene.

Late Pleistocene vegetation was described in the greatest detail by Heusser et al.'s (1999) revision of the Humptulips Bog core originally published in 1964. In the new 7.7-m bog core, analyzed for pollen at 5-cm intervals, the upper 170 cm could be dated by radiocarbon and the lower portion dated only by correlation with hypothesized climatic events. An abbreviated presentation of this pollen diagram, showing ten important pollen types, is shown in Fig. 4.1. This pollen record reveals that the Holocene is distinct from the rest of the core, which suggested to

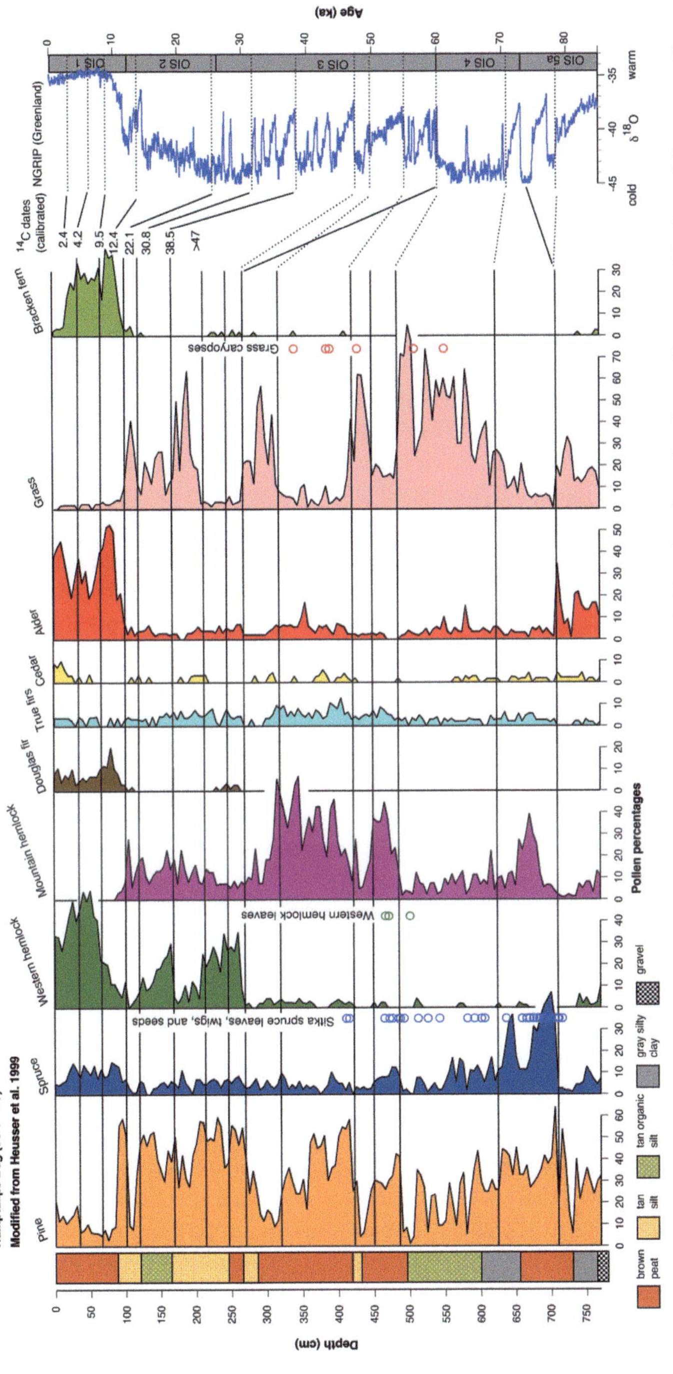

Fig. 4.1 Pollen diagram from Humptulips Bog, redrawn from (Heusser et al. 1999) and abbreviated to a few key pollen types. *Horizontal lines* are pollen zones defined by Heusser et al. The oxygen isotope record from Northern Greenland Ice Core Project (NGRIP) and the marine-based oxygen isotope stages (OIS) are shown for correlation with the pollen record. Solid *sloping lines* indicate age assignments by Heusser et al. (1999) while dashed *sloping lines* are alternate age possibilities based on the timing of interstadials in the NGRIP core. (North Greenland Ice Core Project members 2004)

Heusser et al. (1999) that the base of the core did not reach the previous interglacial. The Holocene vegetation is marked by high values of alder and western hemlock, with consistently high values of Douglas-fir and bracken that suggest regional fire disturbances. A distinct fluctuation in grass and pine at 90 cm depth is consistent with the Younger Dryas event, though radiocarbon ages below that period suggest a hiatus in the core at the time of the LGM. The lower 6.7 m of the core, however, has generally lower alder and greatly fluctuating levels of grass and pine pollen. Heusser et al. (1999) correlated the entire period with low western hemlock and alder pollen and high grass and mountain hemlock to oxygen isotope stage (OIS) 4, which compresses the majority of the core to a 12,000-year period of time. An alternate interpretation is that the units of silt and silty clay correlate to OIS 4 and other fluctuations in grass pollen date to the numerous interstadial (warm) events that occurred in OIS 3 (see dashed lines in Fig. 4.1). Resolving these interpretations will require obtaining new core material and using advanced means of absolute dating such as optically stimulated luminescence on quartz grains.

In addition to the Humptulips Bog, Heusser et al. (1999) analyzed a peat section that underlay several meters of laminated silty lake sediment on the West Fork Humptulips River. This peat section was marked by western hemlock, Sitka spruce, and alder pollen. Based on its infinite radiocarbon age (>50 ka) and the similarity to modern forests, Heusser suggests that it dates to the previous interglacial period (ca. 120 ka, or OIS 5e). Regardless of the details of the Humptulips chronology, this record clearly reveals that for the majority of the Pleistocene the regional vegetation was significantly different from today and fluctuated greatly through time. Heusser et al. (1999) estimated a ≥5 °C decline during the LGM and several times when tree line was close to the site, at 100 m above sea level (asl). Thus, the endemic species believed to have persisted on the peninsula through the Pleistocene occurred in a regional vegetation and climate of mountain-hemlock parkland or tree line, which returned several times during the Pleistocene and persisted for several millennia each time.

4.3   The Late-Glacial and Holocene

Heusser (1973) obtained the greatest detail on the late-glacial and Holocene vegetation of the western Olympic Peninsula from three bog and lake pollen records located just north of the southern limit of the Juan de Fuca Lobe (Heusser 1973). Wessler Bog, located 6.2 km east of the north tip of Ozette Lake, has a basal radiocarbon of its 10-m core at 14,460 ^{14}C yr BP, which calibrates to an age of 17.5 ka. This suggests an initial retreat of ice at the same time or slightly earlier than the retreat of the Vashon Lobe in the Puget Trough (Heusser 1973). Interestingly, many small diameter trees found embedded in glacial till in the same area dated to roughly 13,100 ^{14}C years BP (calibrated to ca. 15.8 ka). Heusser (1973) interpreted this as lodgepole pine trees that quickly colonized on ablation moraines as the ice sheet stagnated for many centuries. Eventually after ca. 15.8 ka, the final loss of the ice resulted in the collapse of moraines and the burial of colonizing forests. While this sequence of events is paralleled in the Puget Trough, the exceptionally long period of stagnating ice inferred from the Olympics needs to be confirmed with better dates constraining the timing of deglaciation.

The pollen records from Wessler Bog, Wentworth Lake, and Soleduc Bog indicate parallel environmental sequences (Heusser 1973). These sites began with cold climates that were characterized by lodgepole pine, followed by the warm early Holocene period with alder and small quantities of Douglas-fir (presently very rare in the region), and then a gradual transition to cool-moist late-successional cedar–hemlock forests. Few radiocarbon dates constrain these pollen records. A more detailed record from Wentworth Lake is presented later in this chapter.

Mathewes and Heusser (1981) extended Heusser's approach of pollen-based temperature reconstructions to another site southwest British Columbia. The statistical method of Heusser et al. (1980) was applied to a higher-resolution 14,000-year pollen record from Marion Lake near Vancouver, British Columbia. These results generally confirmed the previous study, but showed a more distinct "xerothermic" period in the early Holocene that began to become cooler and drier a few centuries before the Mazama ash (7.6 ka).

A distinctive feature of the Holocene pollen records is a transition from fire-disturbed early Holocene forests to forests dominated by shade-tolerant species. This transition to abundant western hemlock and western redcedar was mostly complete by 6 ka, suggesting that this was the first time that old-growth forests were widespread across the Pacific Northwest (Brubaker 1991). Palynologists Rolf Mathewes and Richard Hebda showed that western redcedar increased to modern levels of abundance in a south-to-north pattern, from 8 ka or earlier in southern Washington to 4 ka on northern Vancouver Island (Hebda and Mathewes 1984). As Native American cultures utilize large-diameter western redcedar, the availability of this important resource was likely quite recent in coastal British Columbia.

Other researchers developed detailed pollen records in the Puget Trough in the 1980s (Barnosky 1981; Tsukada et al. 1981; Leopold et al. 1982; Sugita and Tsukada 1982; Cwynar 1987). These records, dated by multiple radiocarbon dates, provide a more detailed ecological interpretation of the late-glacial and Holocene vegetation and climate histories than possible at the time of Hansen's and Heusser's work. A synthesis of pollen records from western Washington (Whitlock 1992) indicated a 1000-m lowering of alpine tree line during the LGM, with humid conditions occurring only on the western Olympic Peninsula. At the time of maximum ice extent of the Puget Lobe, mesic subalpine parkland characterized many of the low-elevation sites both east and west of the Olympic Mountains. Following the retreat of the Vashon Lobe, the arrival of temperate taxa resulted in rapid vegetation changes between 15 and 10 ka, many of which were mediated by the influence of forest fire (Cwynar 1987; Whitlock 1992; Prichard et al. 2009). These vegetation changes are the focus of the next section of this chapter with an emphasis on five postglacial pollen records from the Olympic Peninsula.

4.4   Postglacial Vegetation and Fire at Five Sites on the Olympic Peninsula

The extensive work by Hansen and Heusser remains the best view into the LGM and earlier environments on the peninsula. However, there remain two limitations to many of their studies. First, the radiocarbon dating methods used may have been susceptible to contamination of older carbon from groundwater sources, causing some dates of the sediment matrix to be older than that of the associated pollen record. The overall insufficiency of dates at all sites limits their utility for regional correlation. Second, the pollen of Cupressaceae, the family that contains western redcedar and Alaska yellow cedar, was inconsistently identified. Heusser did indicate these taxa on his diagrams but pollen percentages almost never exceeded 2%; Bogachiel Bog and the Humptulips cores are the exceptions (Heusser 1978; Heusser et al. 1999). Although this pollen type is sometimes difficult to distinguish from algal cysts or other pollen and spore types, palynologists have been able to reliably identify it using modern chemical processing techniques to obtain a cleaner sample for microscopic identification (Bortenschlager 1990).

Beginning in 1993, Linda Brubaker from the University of Washington initiated new paleoecological studies of the Olympic Peninsula. Sediment cores were collected from small lakes located at sites spanning across the precipitation and elevational gradients of eastern and western Olympics, including the first records to be analyzed from montane and subalpine elevations (see Fig. 1.5 for climographs). These records have been published, with the exception of Wentworth Lake, a site originally studied by Heusser (1973). However, these sites have never been considered simultaneously to examine the concurrent changes in vegetation across the range of vegetation zones on the peninsula. In the remainder of this chapter, these pollen records are presented in new diagrams, which are described briefly and the significant findings are highlighted. The vegetation histories of these sites are then summarized using a modern analog method in which the pollen data are compared with modern pollen data to link past vegetation to modern forest zones. In the last section, these sites are synthesized with the existing paleoecological records from the region, especially southern Vancouver Island and the Puget Lowland. The goal of this synthesis is to assess variability in vegetation change across the region, with particular attention to whether the isolation of the peninsula has affected its vegetation history.

4.4.1   Study Sites

Wentworth Lake is a 6-m-deep, 14-ha kettle lake located at 47 m asl, 7 km southeast of the southern tip of Ozette Lake in the northwest Olympic Peninsula. There are no inflowing streams. The terrain around most of the lake is level and the size of the watershed is indeterminate. Aerial photographs indicate fluctuating water levels, with a 3-m drop in water from early to late summer, exposing the NW portion of the lake. The forests around the lake are composed of western redcedar, western hemlock, and Sitka spruce, belonging to the Sitka spruce zone. This site was first studied by Heusser (1973).

Yahoo Lake is a 17.8-m-deep, 3.7-ha cirque lake located at 717 m asl on a ridge above Stequaleho Creek and 5.5 km north of the Queets River. The lake has a watershed of only 9.6 ha and no inflowing streams. The site is located at the lower limits of the Pacific silver fir zone. The forests are composed on western hemlock, western redcedar, Pacific silver fir, and minor amounts of Douglas-fir and Sitka spruce. The record from this site was published in Gavin et al. (2013).

Martins Lake is a 9.6-m-deep, 0.8-ha moraine-dammed lake. It is the larger of two lakes located at 1423 m asl on a ridge in the center of the park on the north side of Mount Christie and south of Low Divide. The watershed is very restricted (ca. 3 ha) and there are no streams inflowing or outflowing. The lake is in subalpine parkland within the Mountain hemlock zone, with small patches of mountain hemlock and Pacific silver fir. Common meadow vegetation is red heather, white mountain heather, American bistort, cascade blueberry, glacier lily, and sedge species. Slide alder occurs in avalanche chutes. The site has been published in Gavin et al. (2001).

Moose Lake is a 7.8-m-deep, 3.7-ha cirque lake located at 1544 m asl in Grand Valley 7 km SE of Hurricane Ridge. Its watershed, ca. 400 ha in size, includes steep slopes extending to above 2000 m asl that feed three intermittent inflowing streams. The watershed is composed of talus, scree, and parkland of subalpine fir belonging to the Subalpine Fir Zone. Less common trees include mountain hemlock, Alaska yellow cedar, and lodgepole pine. Meadows host a high diversity of herbaceous species, with distributions primarily controlled by moisture availability. The site has been published in Gavin et al. (2001) and Brubaker and McLachlan (1996).

Crocker Lake is a 5.1-m-deep, 26-ha kettle lake located at 54 m asl in the Leland Valley in the northwestern peninsula, 6.5 km south of Discovery Bay. The watershed of Crocker Lake includes the area drained by Andrews Creek, estimated to be more than 1000 ha. Forests in the area are heavily managed Douglas-fir stands; red alder and western redcedar were likely important historically and especially along the streams. Despite being in the Western hemlock zone, this late-successional species is rare because of dominance of Douglas-fir after disturbance in this drier climate. The site has been published in McLachlan and Brubaker (1995) and Brubaker and McLachlan (1996).

4.4.2   Field and Laboratory Methods

Sediment cores were taken using a Livingstone piston corer operated from a plywood platform on inflatable rafts that were anchored over the deepest part of each lake (Wright et al. 1984). Cores were wrapped in aluminum foil and transported to the University of Washington. Coring was performed during the summers of 1993 and 1994.

Radiocarbon dates were obtained on bulk sediment using conventional dating or on plant macrofossils using accelerator mass spectrometry (AMS). Two to seven dates were obtained from each site, or 24 dates in total. Radiocarbon dates were calibrated using the INTCAL09 calibration curve, which produces an irregular probability distribution for each date (Reimer et al. 2009). Chronologies were developed using a smoothing spline to the median age of each calibrated radiocarbon date. Bulk radiocarbon dates were assessed for potential contamination by carbonate sources by comparison with independent information on glacial retreat dates or the ages of the Mazama tephra. Chronologies at three published sites (Crocker, Moose, and Martins lakes) were revised for this book.

Sediment organic matter was quantified using loss on ignition (LOI) at 550 °C. Pollen analysis was conducted on 1-cm^3 samples following standard methods (Faegri and Iversen 2000) with the addition of a 7-µm sieve step for basal samples with high clay content (Cwynar et al. 1979). Slide alder and red alder pollen were separated based on pore morphology (Sugita 1990). Pollen percentages were based on a sum of at least 350 terrestrial pollen grains.

Plant macrofossils and charcoal were analyzed in sediment subsamples taken at intervals between 1 and 10 cm. For Wentworth Lake, 54 samples were taken contiguously at an interval of approximately 250 years, corresponding to 7–20-cm intervals in the core. For Yahoo Lake, 415 samples, each 5 cm^3, were taken contiguously at 1-cm intervals. For Martins Lake, 47 samples were taken contiguously at 5-cm intervals. For Moose Lake, 41 subsamples were taken contiguously at 7–25-cm intervals. For Crocker Lake, 26 subsamples each spanning 10–25 cm of core, were taken irregularly down core. For all lakes, sediment subsamples were wet sieved at 0.5 mm for charcoal and 1.18 mm for macrofossil identification. For all sites except Yahoo Lake, sediments were soaked in sodium hexametaphosphate solution prior to sieving. For Yahoo Lake, sediment subsamples were wet sieved at 150 and 1180 µm. The > 1180-µm fraction was saved for macrofossil identification. The 150–1180-µm fraction was treated with a 5 % KOH solution at 40 °C for 20 min and then washed through 500- and 150-µm sieves with a gentle flow of water. The 150–500-µm fraction was saved for charcoal analysis. Plant macrofossils and charcoal particles were tallied under 20–70× magnification. Macrofossils were identified using published keys for conifer foliage (Dunwiddie 1985) and a reference collection at the University of Washington. Needle fragments were combined and expressed as needle equivalents and *Thuja* branchlets were tallied individually following Dunwiddie (1987). Charcoal was expressed as an accumulation rate (charcoal accumulation rates, CHAR; pieces cm^{-2} year^{-1}).

Pollen data from the fossil cores were compared to the modern pollen assemblages on the Olympic Peninsula using the modern analog technique. This method aims to determine, for each fossil pollen assemblage, the closest modern analog to a set of modern pollen assemblages, indicating an analog of vegetation types. To do this, a dissimilarity coefficient is calculated between each fossil pollen assemblage and each modern pollen assemblage. If the coefficient falls below a threshold, then the fossil assemblage is matched with a modern analog. The modern assemblages come from surface sediments from 65 small lakes located in areas of the peninsula where vegetation is little disturbed; heavily logged areas were avoided (Fig. 4.2, Gavin et al. 2005). Heusser (1969) conducted a similar study on the Olympic Peninsula

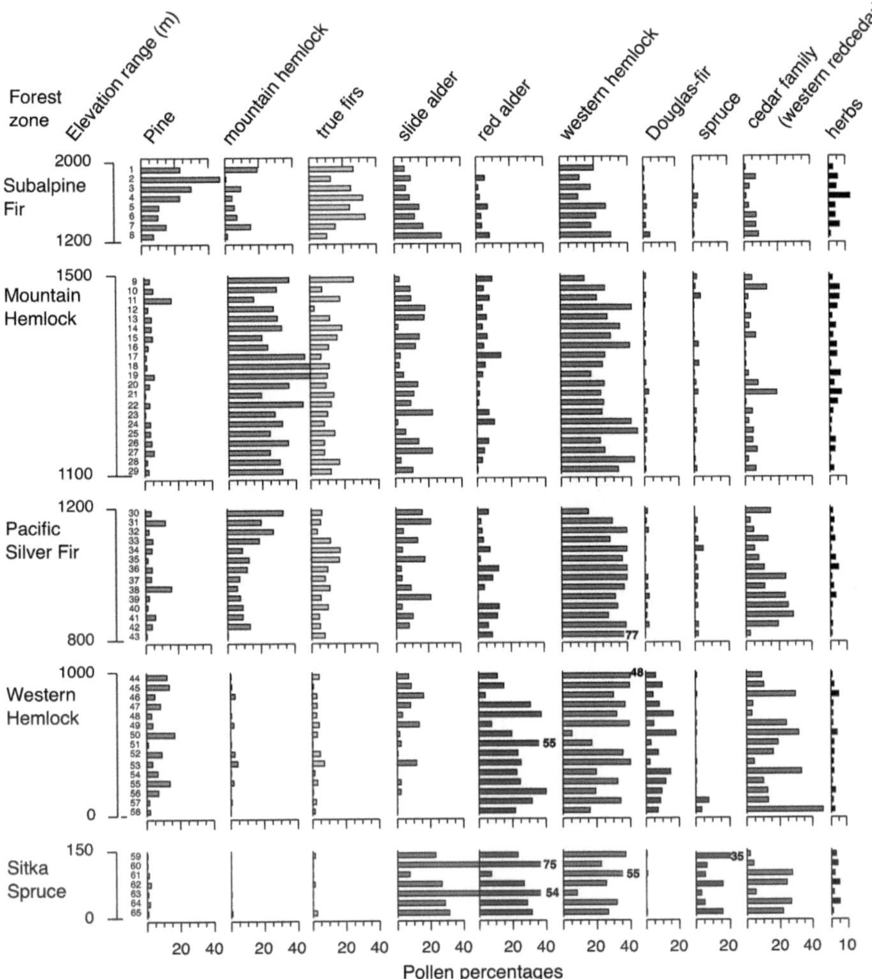

Fig. 4.2 Modern pollen assemblages from 65 small lakes on the Olympic Peninsula. These data were compared to fossil pollen assemblages using the modern analog technique. (Figure modified from Gavin et al. 2005)

but used subcanopy (moss polster) pollen assemblages, which are not adequate for comparing to the lacustrine pollen assemblages used in this study because pollen assemblages from subcanopy locations are characteristically different than those from those of nearby lakes, especially near the tree line (Minckley and Whitlock 2000).

For these comparisons, we used the squared chord distance dissimilarity coefficient, calculated as $D_{m,f} = \sum_{k=1}^{n} \left(\sqrt{p_{m,k}} - \sqrt{p_{f,k}} \right)^2$ where $D_{m,f}$ is the squared chord distance between a modern and fossil assemblage, and $p_{m,k}$ and $p_{f,k}$ are pollen proportions of pollen taxa k in modern and fossil assemblages, respectively, and there are a total of *n* pollen taxa. Gavin et al. (2005) found that distances of 0.15–0.30 occurred among modern assemblages within vegetation zones on the Olympic Peninsula. Those threshold distances could be used as conservative and liberal criteria, respectively, for assessing analogs between fossil and modern assemblages.

4.4.3 Sediment Stratigraphy and Chronology

The 937-cm sediment core from Wentworth Lake is composed of silty clay with occasional sandy layers of 1–3 cm in length below 530 cm (LOI 2–5 %), brown uniform gyttja from 530–490 cm, and organic gyttja above 470 cm (LOI 30–45 %). In the zone above the silty gray clay, the five AMS radiocarbon dates fell on a smooth line, which intersected a very faint but discernable Mazama tephra (Table 4.1, Fig. 4.3). The basal silty clay layer contained no pollen and we believe it was rapidly deposited. No basal radiocarbon date was obtained.

The 430-cm core from Yahoo Lake is composed of uniform light gray silt below 348 cm (LOI 4–8 %) that contained an organic layer in which LOI increased up to 40 %. The upper 348 cm is finely laminated organic sediment (LOI 45–70 %) with the exception of a 1-cm Mazama tephra. The five AMS radiocarbon dates fall on a smooth curve and in line with the Mazama tephra.

The 277-cm core from Martins Lake is composed of uniform silty clay (LOI 2–8 %) in the lower 136 cm and laminated organic gyttja (LOI 13–25 %) in the upper 137 cm; these units are separated by a 4 cm Mazama ash. Two AMS radiocarbon dates fall on a smooth curve in line with the Mazama tephra. Lack of macrofossils for dating the lower portion of the core necessitated correlating the pine pollen decline at Martins with the same feature at Yahoo Lake.

The 708-cm core from Moose Lake is composed of silty clay with sand layers (LOI 4–6 %) below 550 cm, and uniform organic gyttja (LOI 10–25 %) in the upper 550 cm with a 1-cm narrow Mazama tephra. The age model used in Gavin et al. (2001) was revised after considering a likely error in one of the conventional (bulk) radiocarbon dates. The conventional date at 565–575 cm places the distinct decline in pine pollen more than 1000 years earlier than it occurs regionally. A very old age of 30 ka at 695 cm is a good indication that old carbon is contributing to the radiocarbon ages at the base of the core. Assigning an age of 11.6 ka to the pine decline, in contrast, results in a pollen stratigraphy in line with the established regional chronology. We found that the decline of pine pollen to very low levels is complete by 11.6 ka consistently at three sites in the region (Tsukada et al. 1981; Leopold et al. 1982; Brown and Hebda 2002) and agrees with our dates at Yahoo Lake.

The 1181-cm core from Crocker Lake is composed of silty clay with sand layers (LOI 2–13 %) below 930 cm and organic gyttja (LOI 20–40 %) in the uppermost 930 cm, with the exception of a 1-cm Mazama tephra. The age model used in McLachlan and Brubaker (1995) was revised after considering two likely errors in the current set of dates. First, a conventional radiocarbon date 13 cm above the

Table 4.1 Radiocarbon dates on five sediment cores from the Olympic Peninsula

	Depth (cm)	Lab code	14C age ± 1 SD	Material dated or age basis[a]	Calibrated age
Wentworth Lake					
	90–92	AA-20319	3285 ± 65	Conifer needles	3520 (3380–3640)
	183–188	AA-20320	6210 ± 120	Conifer seeds and cone	7100 (6830–7330)
	206–206.5			Mazama tephra[b]	7627 (7477–7777)
	281–285	AA-20321	8540 ± 120	Conifer seeds and alder seed	9530 (9260–9910)
	444–445	CAMS-25018	10,980 ± 160		12,880 (12,590–13,180)
	510–520	AA-20322	11,985 ± 140	Seeds	13,840 (13,460–14,160)
Yahoo Lake					
	109–111	CAMS-35107	3620 ± 60	Conifer needles	3940 (3820–4090)
	198–199	AA-20315	5845 ± 60	Conifer needles	6660 (6500–6790)
	230–231			Mazama tephra[b]	7627 (7477–7777)
	293–295	AA-20316	8175 ± 110	Conifer needles	9140 (8850–9440)
	356–357	AA-20317	9980 ± 115	Conifer twig and wood	11,510 (11,210–11,830)
	407–409	AA-20318	11,915 ± 95	Conifer needles	13,770 (13,480–13,980)
Martins Lake					
	71–73	AA-20323	1995 ± 100	Conifer needles	1960 (1710–2160)
	122–123	CAMS-32689	5390 ± 80	Tsuga mertensiana cone	6180 (5990–6310)
	137–141			Mazama tephra[b]	7627 (7477–7777)
	232		Correlation with regional Pinus decline	11,600	
Moose Lake					
	150–160	Beta-63814	3280 ± 70	Bulk sediment	3510 (3370–3640)
	350–360	Beta-63815	6130 ± 90	Bulk sediment	7020 (6790–7250)
	413–414			Mazama tephra[b]	7627 (7477–7777)
	480–488	AA-20324	6705 ± 70[d]	Woody material	7570 (7460–7680)
	565–575	Beta-63816	11,040 ± 70[c]	Bulk sediment	12,930 (12,710–13,110)
	590		Correlation with regional Pinus decline	11,600	
	695–705	Beta-63817	30,709 ± 1490[c]	Bulk sediment	35,280 (31,900–38,580)

Table 4.1 (continued)

	Depth (cm)	Lab code	14C age±1 SD	Material dated or age basis[a]	Calibrated age
Crocker Lake					
	200–210	Beta-60299	1780±70	Bulk sediment	1700 (1550–1870)
	450–460	Beta-60300	4760±80	Bulk sediment	5490 (5320–5610)
	601–611	Beta-60301	7530±130[c,d]	Bulk sediment	8330 (8150–8560)
	613–614			Mazama tephra[b]	7627 (7477–7777)
	800–810	Beta-60302	9830±140[c]	Bulk sediment	11,280 (10,760–11,770)
	900		Correlation with regional Pinus decline	11,600	
	950–960	Beta-60303	11,540±240[c,e]	Bulk sediment	13,410 (12,900–13,900)
	1174	Beta-70546	10,420±60	Wood	12,300 (12,090–12,530)

[a] Bulk sediment ages are conventional radiocarbon dates, while all other dates on plant material are accelerator mass spectrometry radiocarbon dates

[b] Mazama tephra date is from Zdanowicz et al. (1999)

[c] Date not used due to likely old carbon in bulk sediment

[d] Date not used because not in stratigraphic order with Mazama tephra

[e] Small sample (0.4 gm) with extended running time

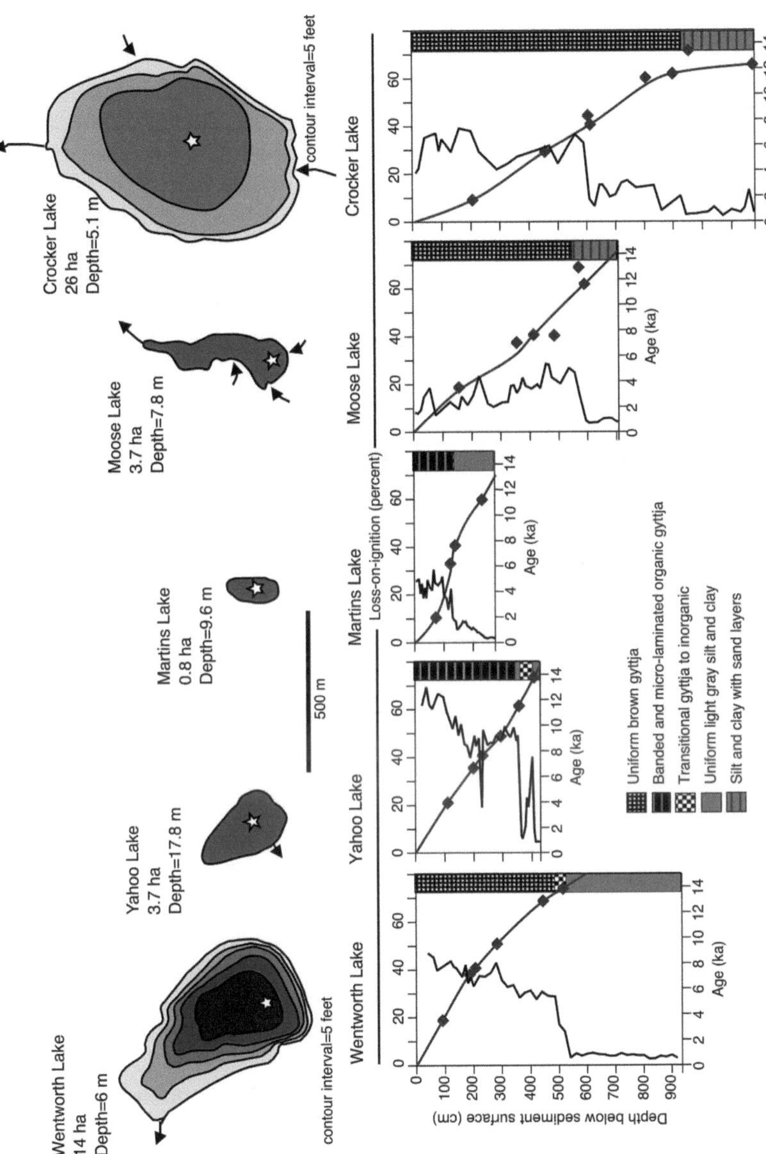

Fig. 4.3 Maps, age–depth relationships, and stratigraphy of sediment cores from five lakes on the Olympic Peninsula. Bathymetric maps shown for two lakes where bathymetric data were available. *Blue diamonds* indicate radiocarbon dates and the *black line* is loss on ignition at 550 °C, a measure of organic matter in lake sediment

Mazama tephra was 800 radiocarbon years older than the known age of the tephra, suggesting that other conventional dates in the core are also too old due to old carbon fixed by algal productivity in the lake (see Brown (1994) for precision dating of the Mazama tephra at Crocker Lake using AMS dates on a pollen extraction). As the conventional dates are considered unreliable, we used a correlation of the decline in pine pollen as used at Moose Lake. Second, an AMS radiocarbon date near the base of the core is ca. 1000 years younger than a bulk-sediment date 1.4 m higher in the core (at 950–960 cm). The bulk-sediment date is likely too old for the reasons given above. We note that the Leland Creek spillway, in which Crocker Lake is located, was actively draining a proglacial lake behind the Vashon lobe 4000 years earlier (at 16.4 ka, Porter and Swanson 1998) than the basal date of the site (12.3 ka). The landscape may have thus supported dead ice and been unstable for millennia before the lake basins formed and organic matter accumulated, such as at the other low-elevation site (Wentworth Lake). A site only a few kilometers away, Cedar Swamp, has a basal AMS date that is 900 years older than Crocker Lake, suggesting variation in when sedimentation began in the kettle lakes of the valley (McLachlan and Brubaker 1995).

Pollen diagrams are plotted with macrofossil occurrences overlaid on the pollen percentages (Figs. 4.4, 4.5, 4.6, 4.7, 4.8) and the modern analog analysis results are plotted for all sites together (Fig. 4.9).

4.4.4 Late-Glacial: 14–11.6 ka

The environment during the several millennia following deglaciation was very dynamic. Climatic fluctuations were pronounced when the waning Cordilleran Ice Sheet and Laurentide Ice Sheet were shrinking in size. A landscape of recently deposited till eroded rapidly before the last stagnant ice melted and the vegetation could stabilize soils. The process of landscape stabilization, though often considered to be very rapid in the Pacific Northwest (e.g., Whitlock 1992), likely varied over space depending on the timing of deglaciation, topography, and the nature of the till deposits. All pollen records indicate that lodgepole pine was the dominant species colonizing the landscape. Today, the distribution of lodgepole pine is restricted, occurring either in early successional forests in the Subalpine fir zone or in coastal populations restricted to bluffs and muskegs.

Forests first appeared on the western Olympic Mountains shortly after deglaciation of the lake sites. Prior to 14.2 ka, tree line, as indicated by low tree but high grass, sedge, and sagebrush pollen (Fig. 4.5), was below 700 m on western Olympics. No other pollen record exists at the same time at lower elevations. Stagnating ice during the retreat of the Juan de Fuca lobe may have precluded a pollen record at Wentworth Lake. Heusser's sites (Heusser 1973, 1974; Heusser et al. 1999) to the south extend further back in time, but resolve this time period poorly. After a decline in pine at 13.6 ka at both Wentworth and Yahoo lakes, a Sitka spruce–western hemlock forest established at Wentworth Lake and a diverse subalpine for-

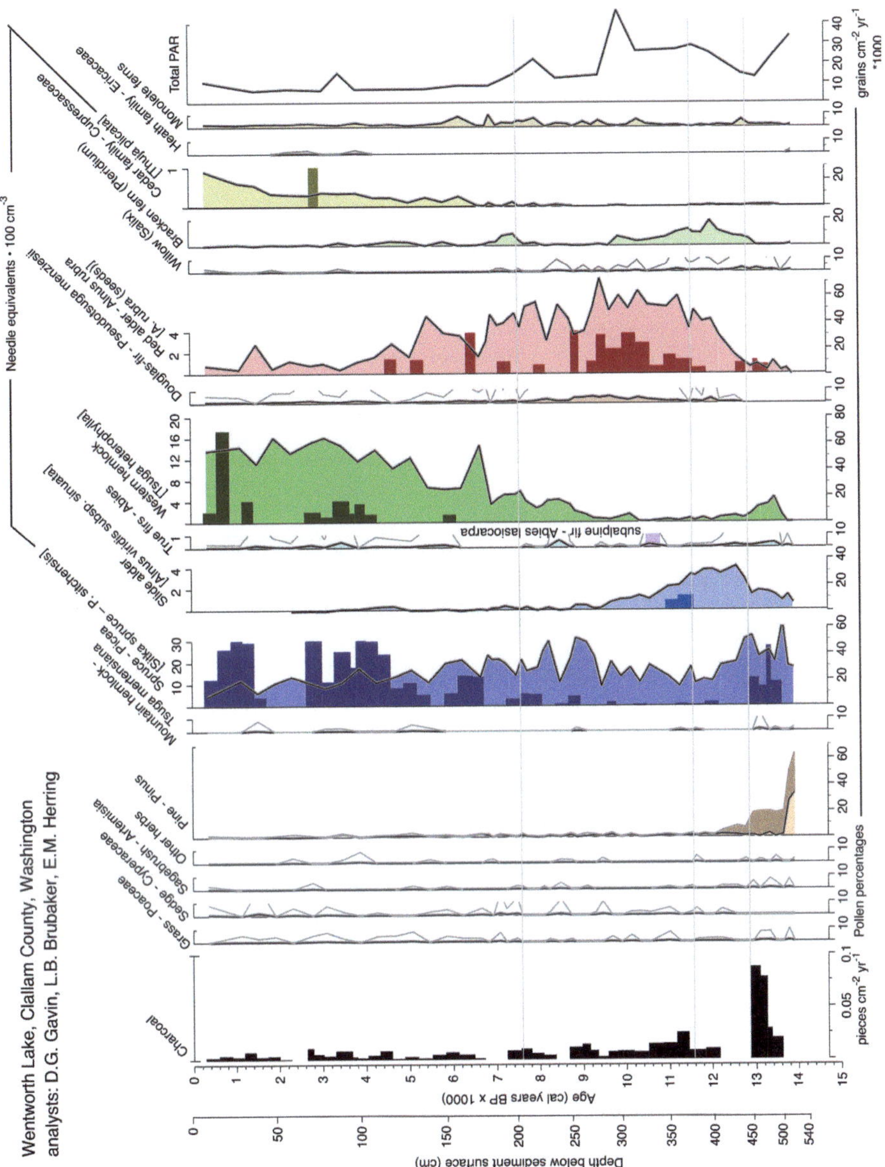

Fig. 4.4 Pollen and macrofossil diagram for Wentworth Lake. Each pollen taxon is labeled above each graph in both common and scientific names. Taxon names in brackets refer to the macrofossil data, graphed as bar charts superimposed over the filled line plot for pollen percentages. *Horizontal lines* indicate the ages of the Mazama tephra (7.6 ka), and the start and end of the Younger Dryas (12.9–11.6 ka). *Gray lines* on the pollen graphs indicate 10× exaggeration for rare pollen taxa. *Pinus* pollen is divided into subgenus Pinus (*light brown*) and undifferentiated *Pinus* (*dark brown*). PAR=pollen accumulation rate

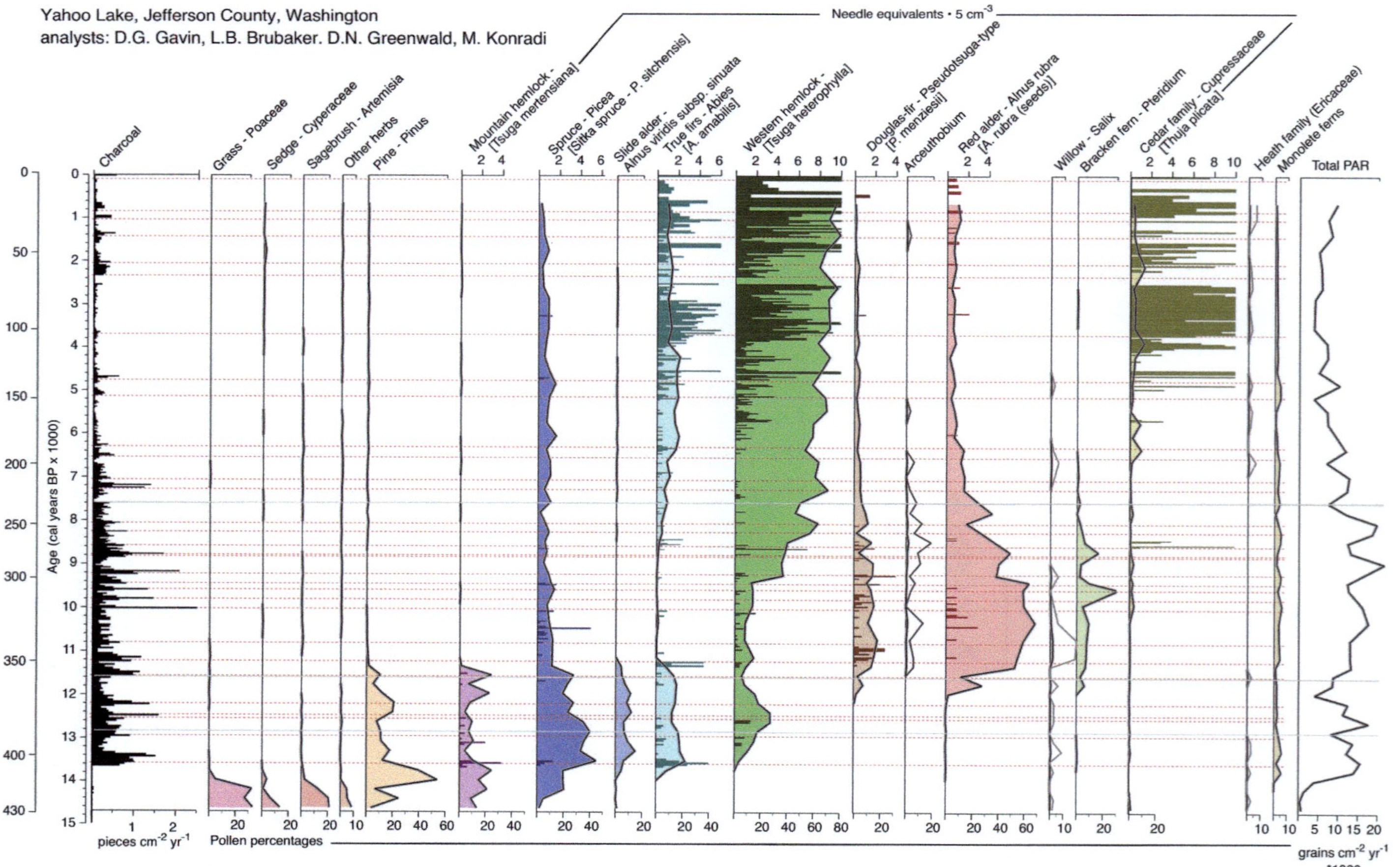

Fig. 4.5 Pollen and macrofossil diagram for Yahoo Lake (Gavin et al. 2013). The diagram is constructed as for Wentworth Lake. Identified fire events from the 1-cm charcoal stratigraphy are shown by dashed red lines. Methods for identifying fire events are described in Higuera et al. (2010)

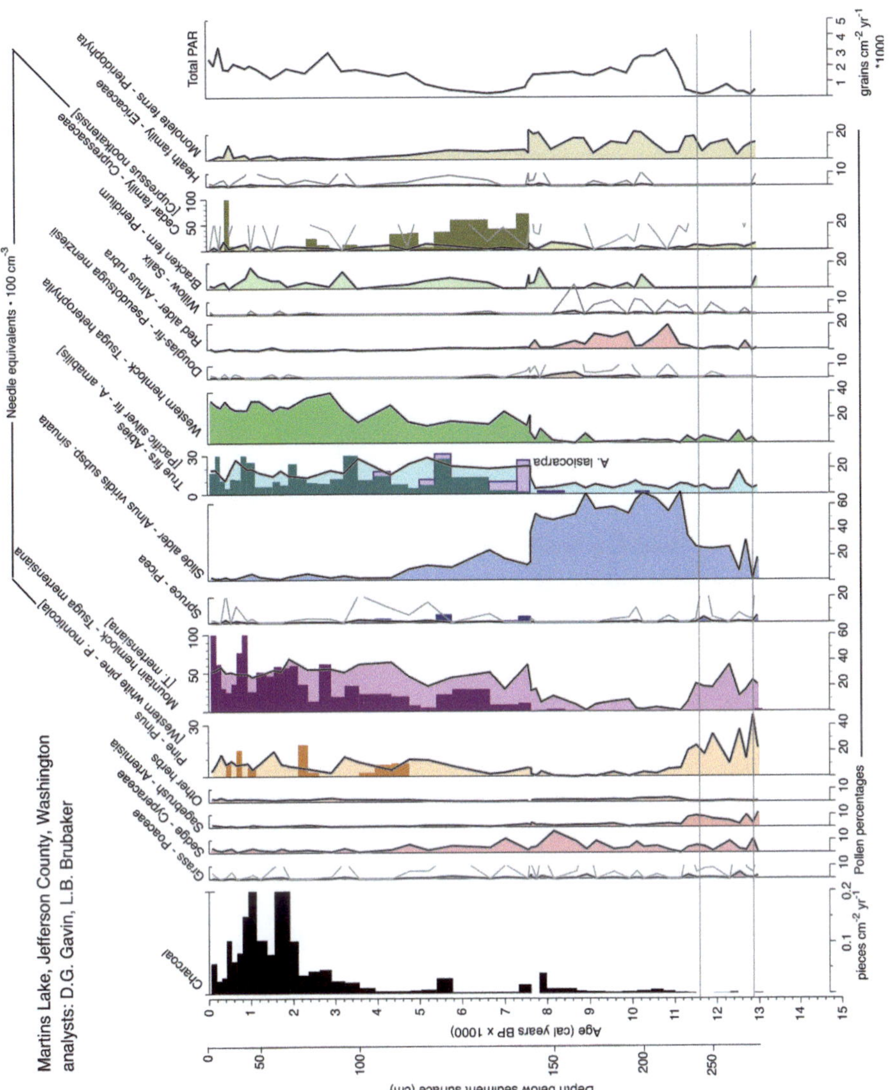

Fig. 4.6 Pollen and macrofossil diagram for Martins Lake (Gavin et al. 2001). The diagram is constructed as for Wentworth Lake

Unlike Wentworth and Yahoo lakes, the high elevation windward and rain shadow sites (Martins and Moose lakes) show very low forest density during the lateglacial. Although the pollen assemblages at these sites are similar to the modern Subalpine fir zone on the peninsula today, tree cover was notably absent unlike today's parkland and meadow vegetation. For example, although Martins Lake had high pine and alder percentages (both regionally dispersed pollen taxa), very sparse

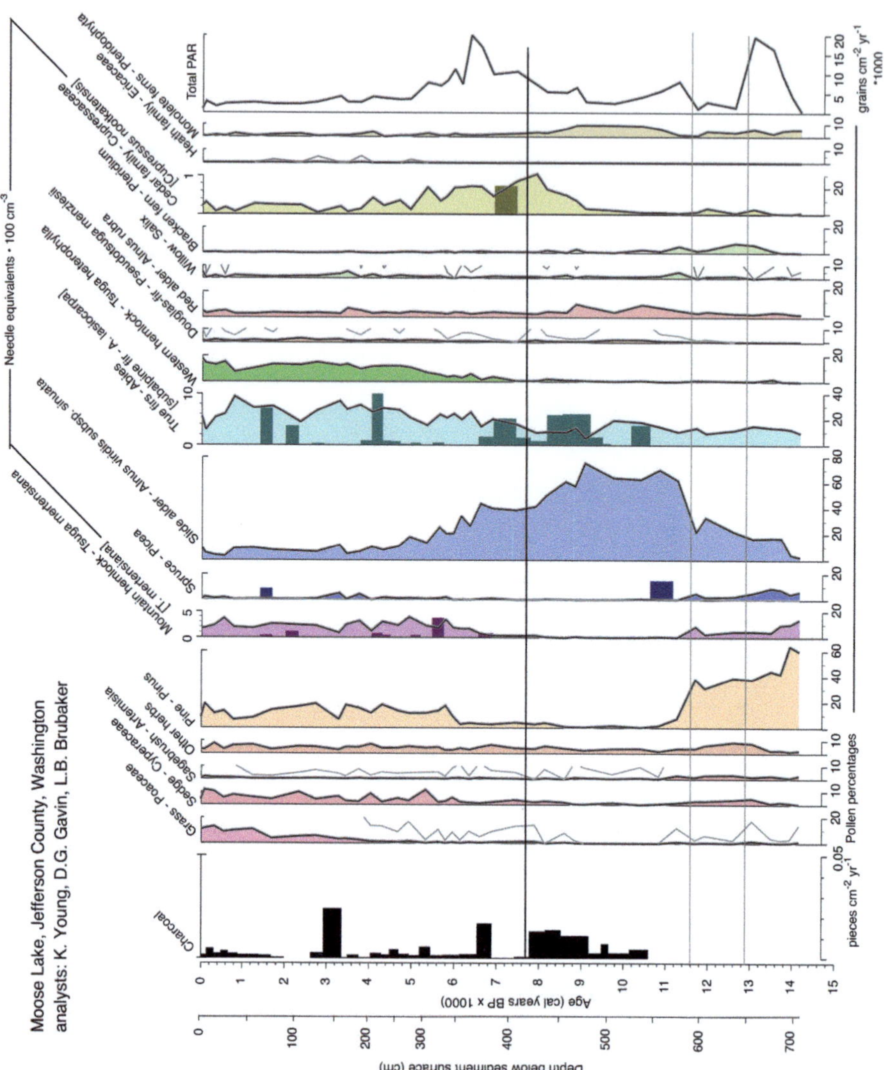

Fig. 4.7 Pollen and macrofossil diagram for Moose Lake (Gavin et al. 2001). The diagram is constructed as for Wentworth Lake

est (mountain hemlock, Pacific silver fir, western hemlock and occasional Sitka spruce) established at Yahoo. At Yahoo Lake, distinct charcoal peaks indicate that intense fires occurred at least six times before 11.6 ka. The coarser Wentworth Lake charcoal record indicates that the highest CHAR values occurred at 13.0–13.6 ka. The combination of species at both of these sites is a poor analog to any modern forest on the Olympic Peninsula (Fig. 4.9).

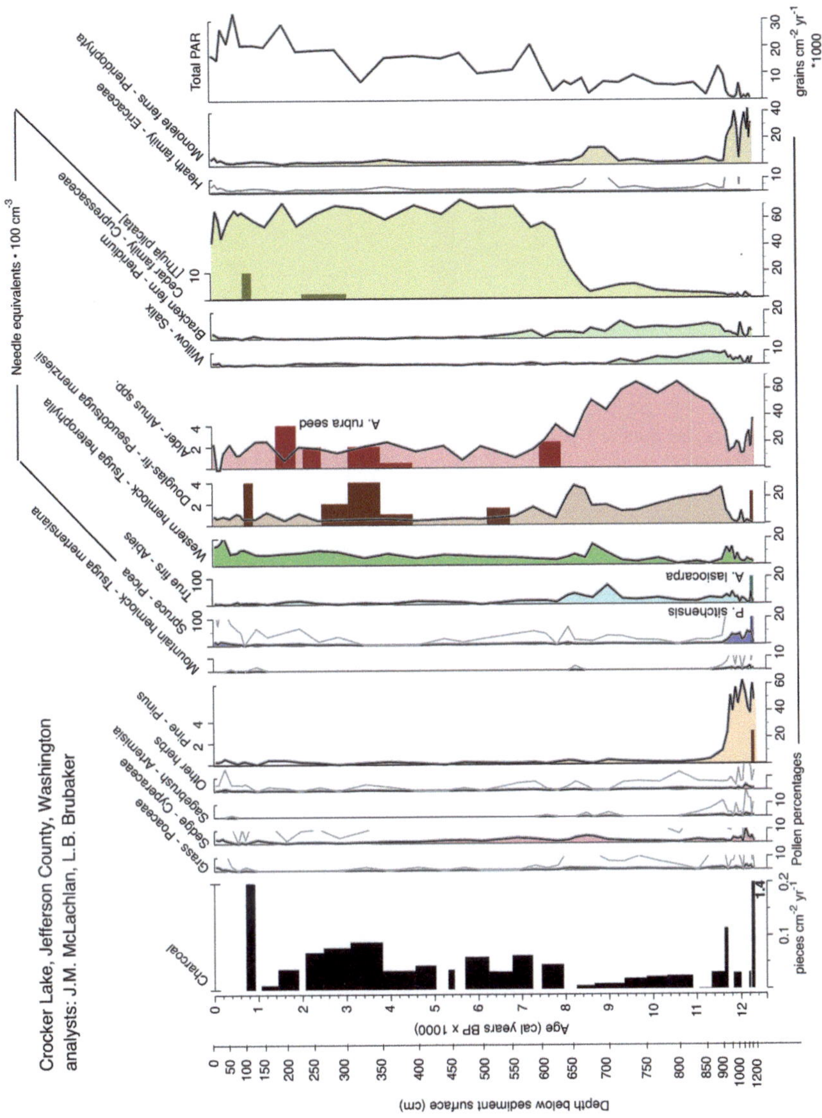

Fig. 4.8 Pollen and macrofossil diagram for Crocker Lake (McLachlan and Brubaker 1995). The diagram is constructed as for Wentworth Lake

tree cover is indicated by very low pollen accumulation, low organic matter, and absence of charcoal and macrofossils. Moose Lake also had high pine percentages, but the lack of any macrofossil or charcoal in the core also indicates sparse or no tree cover.

The deepest sediments of Crocker Lake, in the low-elevation rain shadow of the Olympic Mountains, contained a diverse macrofossil assemblage including Sitka

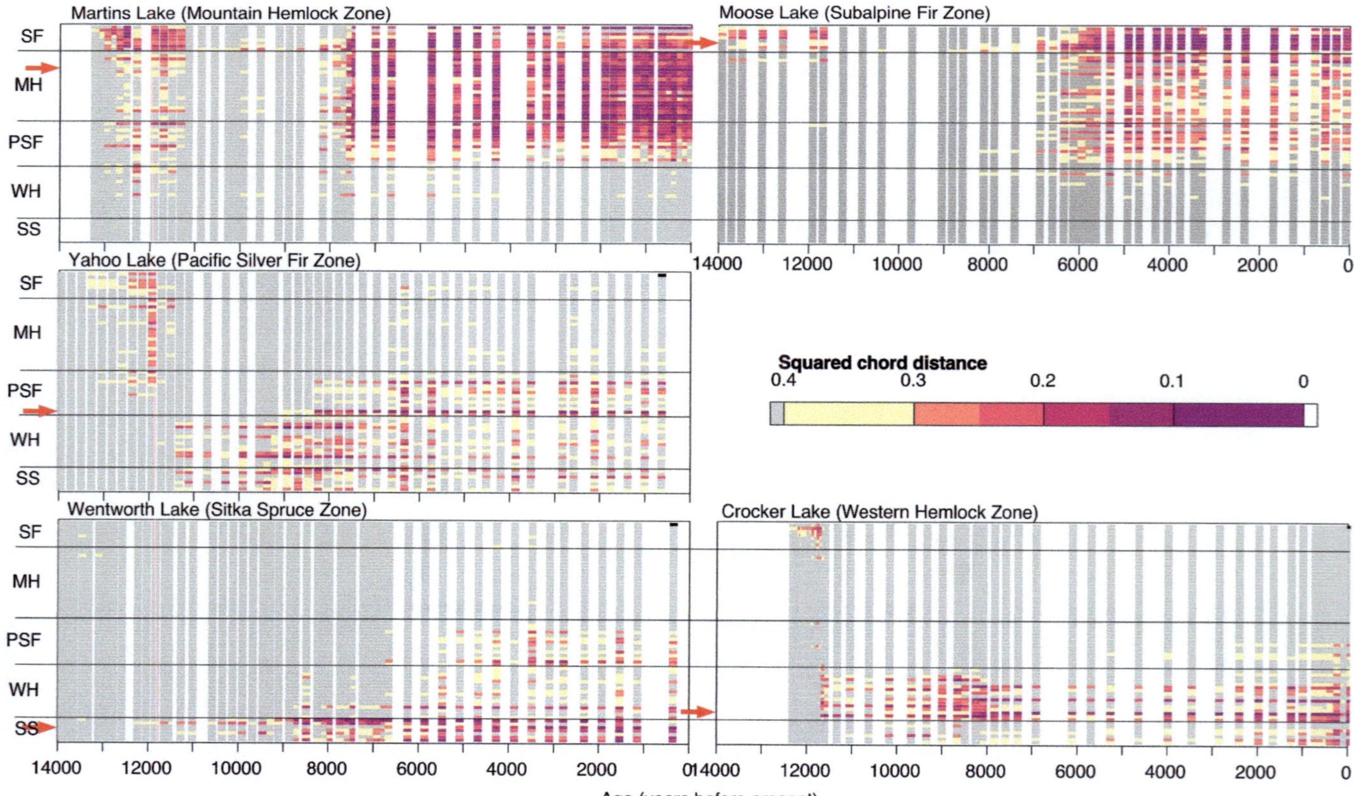

Fig. 4.9 Modern analog analysis for five pollen records on the Olympic Peninsula. For each lake, each column indicates the value of a dissimilarity coefficient (SCD) between the 65 modern pollen assemblages (arrayed by vegetation zone) with a fossil pollen assemblage

spruce, subalpine fir, Douglas-fir, and lodgepole pine needles.[1] The age of these sediments date to within the Younger Dryas interval. The macrofossil sample comes from a 50-cm subsample that may integrate a considerable span of time within the Younger Dryas period (ca. 12.3 ka). These needles and the high charcoal at the base of the core suggest soil developed on the stagnating ice landscape, which later became submerged in the lake basin. Low pollen accumulation rates and the occurrence of herbaceous pollen and fern spores (up to 40%) suggest very open vegetation, which included a minor component of slide alder. Overall, the surprising combination of species (i.e., subalpine fir and Sitka spruce) indicates a vegetation cover that is not analogous to modern zones.

Additional pollen records support this picture for low elevations of the eastern Olympic Peninsula. The nearby Manis Mastodon site located in the center of the rain shadow at Sequim indicates open vegetation, including cactus, from 13.8 to 12.9 ka (Petersen et al. 1983). On Vancouver Island, Brown and Hebda (2003) found a herbaceous-dominated vegetation prior to 14.6 ka followed by a lodgepole pine woodland, as found at Yahoo Lake. The latter period of lodgepole pine woodland or parkland coincides with the widespread retreat of cirque glaciers throughout the Pacific Northwest.

The vegetation responses to the climate changes of the Younger Dryas Chronozone (YDC, 12.9–11.6 ka) vary among the sites. Because paleoclimate proxies from both southern Oregon and British Columbia reveal the YDC as a distinct cold event, the Olympic Peninsula likely experienced this event (Fig. 3.4). Yahoo Lake reveals the YDC most clearly as a period of low organic matter, decreased pollen influx, and few charcoal peaks. During the YDC, Wentworth Lake had few macrofossils and low charcoal, and Moose Lake had a low pollen accumulation rate. The balance of evidence for the YDC suggests a colder and perhaps drier climate that caused reduced vegetation biomass and fewer fires (Gavin et al. 2013). The major vegetation change at the end of the YDC, a decline in pine, may have been locally influenced as soils stabilized and began to support mesic forests. However, dating control at this time is poor at Moose, Martins, and Crocker lakes. Improved radiocarbon chronologies are needed to resolve how this vegetation transition varied spatially.

A synthesis of the sites indicates that as today a very strong rainfall gradient occurred during the late-glacial (Fig. 4.10). On the western peninsula, the occurrence of mountain hemlock pollen at mid-elevations, but not higher, suggests a depressed tree line while at Crocker Lake, on the east side, there was higher pine pollen and indicators of open parkland vegetation. All sites except Wentworth Lake showed a similarity, for at least part of the late-glacial, to the Subalpine fir zone (Fig. 4.9). However, the high-elevation sites were likely in a treeless tundra or shrub tundra and received regional tree pollen from lower elevation.

---

[1] The presence of the Douglas-fir and lodgepole pine needles, not published in McLachlan and Brubaker (1995), was identified by Gavin from archived material.

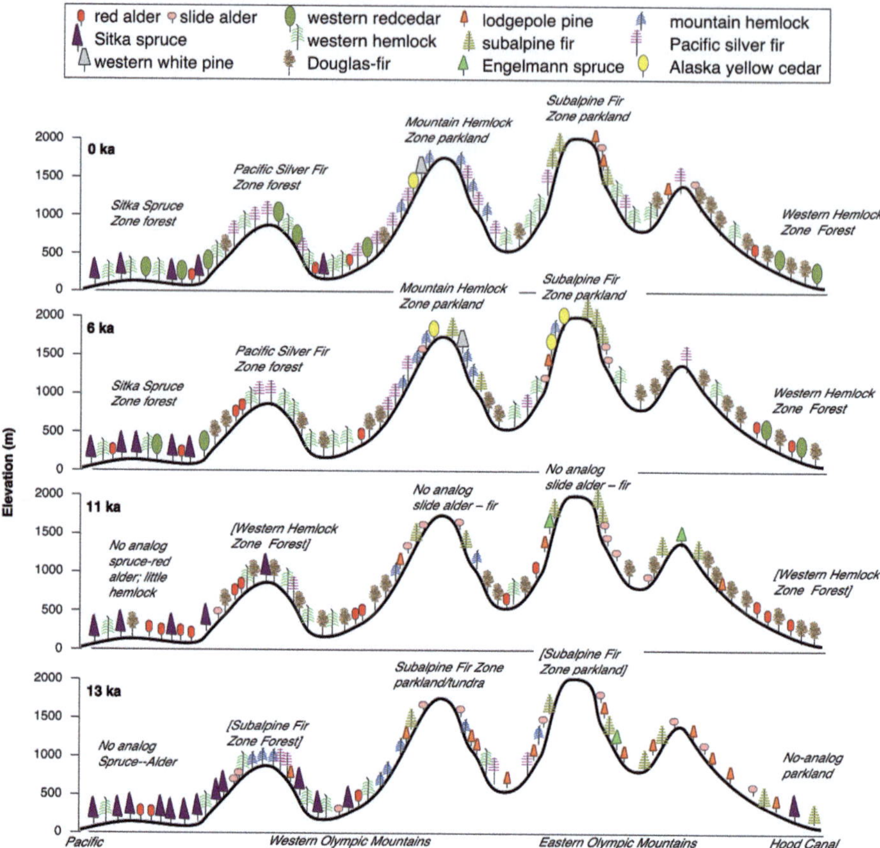

Fig. 4.10 Schematic diagram showing a profile of dominant tree species on a longitudinal transect across the Olympic Peninsula for four time periods. The analog assignments are given above the location of each pollen record along the transect (west to east: Wentworth, Yahoo, Martins, Moose, and Crocker lakes). *Brackets* indicate only marginal analogs were found

4.4.5 Early Holocene: 11.6–6 ka

The start of the Holocene is marked by a great increase in warmth and dramatic changes in forest composition across the Olympic Peninsula. Yahoo Lake exemplifies the rapid turnover in forest composition. At 12 ka, the site supported a diverse mix of subalpine and low-elevation species: Sitka spruce, western hemlock, mountain hemlock, and Pacific silver fir. This assemblage is similar to what occurs in southeast Alaska today, except for presence of Pacific silver fir. Douglas-fir pollen appears first shortly after 12 ka, and a fire (revealed by a sharp charcoal peak) at 11.6 ka resulted in the loss of mountain hemlock and pine and a decline in spruce. However, the macrofossil record indicates that it was not until a second fire at 11.2 ka that Pacific silver fir was lost from the forest and replaced by Douglas-fir.

Red alder dramatically increases at the beginning of the Holocene, and after 11.2 ka the landscape is characterized by frequent fires with disturbance-adapted Douglas-fir and red alder and a minor component of western hemlock. One fire at 10.8 ka resulted in a succession to Sitka spruce dominance at the site. The return to Sitka spruce is unusual, given modern successional patterns in Douglas-fir forests. However, Douglas-fir returned to the site during a period of frequent fires from 10 to 9 ka. Analog analyses indicate that the forest at this time was similar to modern low-elevation forests on the east side of the peninsula. The occurrence of bracken fern and mistletoe pollen (likely on western hemlock) is consistent with the presence of open forests that are conducive to the spread and growth of these species. The return to mesic forest species began at 9.5 ka with a distinct increase in western hemlock, followed by the return of abundant Pacific silver fir at 8.6 ka and the decline of Douglas-fir after 8 ka. At this time, the "analog" forest type also moved from Western hemlock zone sites to the vegetation currently at Yahoo Lake (Fig. 4.9).

The two low-elevation sites, on eastern and western sides of the peninsula, show a similar pattern to that of Yahoo Lake. At Crocker Lake in the east, the dominance of Douglas-fir and red alder is more pronounced than at Wentworth Lake in the west. Western redcedar, the important mesic-site species, increases in abundance at 8 ka at Crocker Lake. Surprisingly, charcoal was less abundant at Crocker Lake during the early than late Holocene. The large size of Crocker Lake might affect transport of macroscopic charcoal to the site, but the reason for the low charcoal content during the early Holocene is not well understood. At Wentworth Lake in the west, red alder and Douglas-fir also increase but Sitka spruce remained abundant and Douglas-fir was a minor component compared to Crocker Lake. As at Yahoo Lake, charcoal was more abundant than later in the Holocene. Red alder and Sitka spruce appear to have been the most successful species to establish after fire, resulting in a vegetation type with no analog to modern forest types on the peninsula (Fig. 4.9). Underscoring the no-analog vegetation at Wentworth Lake, a subalpine fir needle was found at 10.5 ka during the period of high red alder. To the authors' knowledge, the nearest location of subalpine fir to Wentworth Lake is in subalpine forest on a dry south-facing slope above Lake Crescent, 57 km to the east. Although fir pollen was never abundant at this site, we must acknowledge the possibility that subalpine fir established during the late-glacial on well-drained glacial moraines near the lake and survived there for several thousand years into the warm and dry early Holocene. While a combination of Sitka spruce and subalpine fir is nonsensical from the perspective of modern vegetation gradients on the Olympic Peninsula, these species do comingle in fjordlands of southeast Alaska today and they were also detected together in the late-glacial period at Crocker Lake (see above). The alternative explanation, that subalpine fir needles were dispersed westward from high-elevation forests, is very improbable, though Pisaric (2002) found that convection from a forest fire in Montana transported subalpine fir needles up to 20 km.

The early Holocene vegetation at Martins and Moose lakes, located in the current subalpine zone, was unexpected based on regional climatic conditions at that time. The warm summer temperature of the early Holocene, up to 4 °C warmer than present, would suggest that tree line should have been higher than at present and that forest density should be greater than in the present parkland vegetation, as has been shown elsewhere in the Pacific Northwest (Rochefort et al. 1994). The core at Martins Lake suggests that trees were extremely rare, as the sediment remained inorganic, the charcoal levels remained very low, the pollen was dominated by regionally dispersed slide alder, and only three small macrofossils of subalpine fir needles occurred in 10.5–7.6 ka. This evidence that forests did not extend above Martins Lake during the early Holocene presents a conundrum. As indicated by modern tree ring studies in the Olympics, increased temperature is expected to favor tree growth in mountain hemlock forests (Zolbrod and Peterson 1999; Halofsky et al. 2011). Gavin et al. (2001) speculated that snow avalanching during the early Holocene was more common than today, due to the colder winter temperatures, greater snow accumulation, followed by rapid warming that may have produced an unstable snowpack. Frequent disturbances from snow avalanches would have favored slide alder thickets over conifer trees. Similarly, slide alder pollen dominates at Moose Lake. However, at this site, the first tree macrofossil (Engelmann spruce) occurs at the start of the Holocene and subalpine fir needles indicate that trees occurred locally throughout the early Holocene. The Moose Lake record offers a striking contrast to Martins, where trees did not become common until directly after the Mazama tephra deposition, when subalpine fir (currently absent from the area), mountain hemlock, and Alaska yellow cedar were established. One possibility is that the Mazama ash, though only 2 cm in thickness in the core, was sufficiently thick on land to fill voids in talus slopes and thus provide a mulch for conifer seedlings to survive the summer drought (Gavin et al. 2001). In line with this hypothesis is that the most drought-adapted conifer, subalpine fir, was the only conifer present before the Mazama ash and was the one species that dominated for the period directly after the ash deposition. This species was replaced by mesic-adapted Pacific silver fir later in the Holocene. The ash deposition did not have as great an effect at Moose Lake, perhaps because, while Martins Lake is on a ridge, Moose Lake is located on a valley floor that received fine-grained material from upslope areas that was suitable for seedling germination even during the warmer early Holocene. Alaska yellow cedar increased in abundance at Moose Lake at the time of the Mazama tephra; the climatic or local edaphic cause of this increase is not clear.

While the distinct early Holocene forests at low elevation are similar to those described by earlier studies in the Puget Lowland and Fraser River valley (Mathewes 1985; Whitlock 1992), this set of five records provides new detail plus some unexpected findings. New details emerged from the one site where a high-resolution macrofossil record was possible (Yahoo Lake). At this site, the turnover of long-lived trees often lagged behind climate change and required major disturbances (Gavin et al. 2013), a prediction emerging from theories and models of forest dynamics (Davis 1986; Kitzberger et al. 2012). The vegetation changes at Yahoo Lake in the early Holocene, such as upslope movement of Sitka spruce, are supported by another detailed macrofossil record near Vancouver BC (Wainman and Mathewes 1987). The vegetation changes in the subalpine, however, were not predicted by expectations of increased tree line under warmer summer conditions.

The controls on tree line in the Olympic Mountains today are complex and vary greatly over space (Woodward et al. 1995; Peterson 1998). Without more paleoecological records, the pattern of subalpine forest development and the role of soils and climate on forest colonization of the subalpine landscape remain unresolved.

4.4.6   Middle and Late Holocene: 6–0 ka

During the middle and late Holocene, cooler and moister summers resulted in lower fire occurrence and the establishment of the dense, deep-shade-tolerant vegetation currently typical in western Washington. At all sites, modern vegetation types were achieved by 6 ka, and the forest composition changed only slightly thereafter (Fig. 4.9). The period around 6 ka marks the time when forests developed their modern (presettlement) species composition and structure. Crocker Lake, for example, began to resemble today's old growth Douglas-fir forests (Brubaker and McLachlan 1996) and Wentworth Lake reached modern levels of shade-tolerant western hemlock. This change to modern conditions is also clearly seen at Yahoo Lake where very long fire intervals (>500 year) established and late-successional species (western hemlock, Pacific silver fir, and redcedar) increased in abundance. At the high-elevation sites, mountain hemlock pollen reached modern levels shortly after 6 ka and slide alder pollen declines to low levels, consistent with increased summer moisture, less snow avalanching, and development of deep organic soils.

Paleoclimate records indicate that the main climatic fluctuations within the past 6000 years were associated with neoglacial advances (Fig. 3.4). While not extensively studied, neoglacial advances on the Olympic Peninsula likely correlate well with other sites in the Pacific Northwest (Long 1975; Hellwig 2010). However, these climate fluctuations were not sufficiently large to affect broad-scale vegetation patterns. The long life spans of the late-successional trees that dominated this period buffered forest compositional change to the decadal and to centennial climate fluctuations of the late Holocene. One exception to the apparent long-term climatic stability of this period is evidenced by an increase in fire at 2.5 ka at Yahoo and Wentworth lakes, which is seen in the Cascade Range as well (Gavin et al. 2007; Prichard et al. 2009). Whether this fire increase is the result of increased climatic variability (favoring intermittent fire-susceptible conditions) or increased burning by humans (for example, for berries and hunting habitat) has been widely speculated and will be further discussed in Chap. 5 (Hallett et al. 2003; Lepofsky et al. 2005).

Other vegetation changes occurring during the late Holocene may represent autogenic ecosystem development rather than direct response to climate fluctuations. The clearest expression of this change is at Wentworth Lake where western redcedar increases progressively through the late Holocene (Fig. 4.4), similar to that found at Whyac Lake on western Vancouver Island (Brown and Hebda 2002) and sites near Vancouver (Wainman and Mathewes 1987). This slow change likely represents the development of redcedar-dominated muskeg forests on the narrow coastal strip due to accumulation of organic matter and increased soil acidity and moisture, both of which favors redcedar. A similar pattern of ecosystem development has been found in coastal forests in southeast Alaska (Hansen and Engstrom 1996). Further evidence of autogenic succession come from a from a wetland site near Crocker Lake where pollen revealed a sequence of hydrosere succession controlled by infilling of the basin with sediment (McLachlan and Brubaker 1995).

4.4.7 Additional Late Holocene Paleovegetation Studies

In contrast to regional pollen records that do not reveal major vegetation change during the late Holocene, pollen records from smaller basins and soils reveal dynamic vegetation at smaller spatial scales. Pollen records from "small hollows" (small bedrock or soil depressions only ca. 5 m in diameter) record vegetation at the stand scale, within about 50 m. Similarly, meadow soils record pollen from primarily insect-pollinated species in much greater proportion than they occur in small lakes. The following three studies of small hollows and soils were conducted within or near the park and document vegetation changes not recorded in larger pollen-collecting sites.

Three small hollow sites were studied on terraces of the Queets River in the western Olympic Peninsula (Greenwald and Brubaker 2001). Two sites within 800 m of the river date only 500 years ago when side channels were abandoned, or a major flood affected the area. These sites show a general succession from red alder to Sitka spruce and western hemlock that is typical following disturbance in this area. The third site, farthest from the river and adjacent to the valley wall, recorded vegetation over the past 5000 years. It showed a consistently high abundance of red alder, which suggests the long-term presence of riparian forests with frequent flooding on the Queets River floodplain. The charcoal record indicated only two fires over the past 5000 years. The only long-term change was an increase in western hemlock over the past 1000 years, which may reflect cooling associated with the Little Ice Age (Greenwald and Brubaker 2001).

In the rain shadow of the Olympic Mountains, a study of small hollows on Orcas Island, revealed histories from 7 ka to present (Sugimura et al. 2008). These records showed that the rocky outcrop of Mount Constitution supported lodgepole pine forests for most of the late Holocene, likely due to the poor soil conditions that support a positive feedback between vegetation and fire. This feedback occurs because lodgepole pine is well adapted to the xeric soil conditions, and lodgepole pine itself is a pyrogenic vegetation type with ladder fuels that result in crown fires. Lodgepole reseeds readily from serotinous cones that open after the fire, completing the fire-vegetation feedback. However, a period from 5.3 to 2 ka supported reduced fire and less pine at sites that had slightly more soil and thus ability to hold moisture during wetter periods (Sugimura et al. 2008).

The only pollen record of small-scale vegetation change at high elevation on the Olympic Peninsula comes from a study by Gavin and Brubaker (1999) of three soil profiles from a ridge top at 1770 m elevation in Royal Basin in the northeastern portion of the park. Subalpine meadows contain high plant diversity within small regions, but these meadows are "invisible" to any pollen records from lake sites. While the 6000-year record from these soil sites was more difficult to study than lake sites due to poor pollen preservation and low temporal resolution, the study revealed that late-snowmelt meadow communities dominated by *Carex nigricans* were the most sensitive to climate change. On these sites, vegetation apparently switched between the monodominant sedge community and more diverse mesic meadow types during the past 6000 years. Other meadow sites revealed a medieval warm period dominated by American bistort (Gavin and Brubaker 1999).

4.5   Regional Synthesis of Vegetation Changes in the Pacific Northwest

Single pollen diagrams, as described above, provide a wealth of information regarding the progression of vegetation and climate changes near a site. However, to examine questions about broad climatic or biogeographic controls of vegetation change at a site, the spatial patterns of vegetation change must be interpreted from synthesizing the information from many pollen diagrams. For example, it is broadly accepted that migration lags have played only a minor role in the postglacial vegetation history of the Pacific Northwest (Whitlock 1992). According to this argument, the heterogeneity of vegetation over mountains ensured that suitable habitat for particular tree species was never far away, and that trees were not displaced very far south during the LGM. Another major benefit of regional synthesis is to understand when modern spatial gradients of vegetation types arose (Blois et al. 2013). Such gradients may have arisen slowly due to limited dispersal, due to competition "sorting out" species along gradients over millennia, or due to changing relative importance of climate variables (e.g., precipitation and temperature). Clearly, the interpretation of the space–time pattern of vegetation change is complex and will not be thoroughly analyzed here. However, other than an early effort (Tattersall 1999) no study from the Pacific Northwest has performed the first step of this analysis, i.e., mapping the pollen records to visualize changing regional patterns in vegetation.

We compiled data from 21 pollen records from western Washington and southwest British Columbia (Fig. 4.11). We used a multistep process to produce the mapped pollen data. We first reanalyzed the data from each site to ensure consistency in interpretation. Appendix A contains detailed criteria for site inclusion, developing age models, and calculation of pollen percentages. We did not include a marine pollen record from Saanich Inlet on Vancouver Island (Pellatt et al. 2001) because its pollen likely represents a broader source area than at lacustrine sites. We then produced maps of pollen abundance for the major tree species at roughly 1000-year time slices from 17 ka (the time of the maximum Cordilleran ice extent) to present. Despite these quality-control efforts, persistent limitations in the accu-

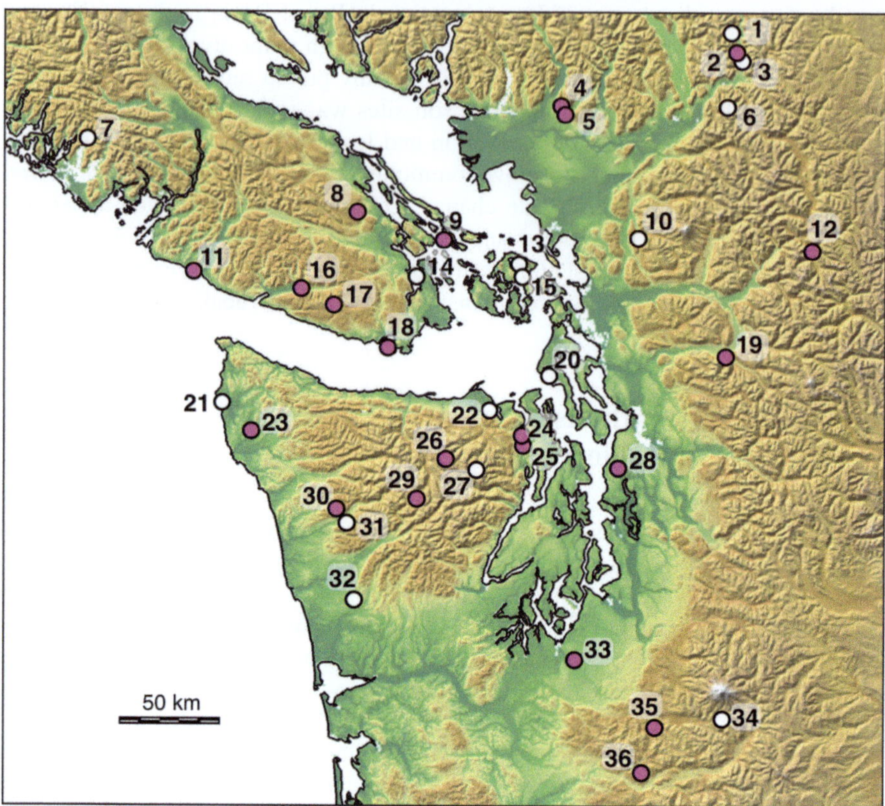

Fig. 4.11 Map of fossil record localities discussed in the text. *Red sites* are radiocarbon-dated pollen records that are mapped in the regional vegetation syntheses

racy of the radiocarbon chronology at many sites preclude a detailed interpretation of individual between-site differences. Furthermore, because sites occur across a range of elevations and microsite types, nearby sites may have very different pollen percentages.

The maps reveal regional patterns of vegetation change, with no two species (or species groups) showing the same pattern. Below, we provide both species-level interpretation and regional vegetation interpretations, invoking biogeographic controls such as dispersal limitation, competition, and climatic constraints on species. The pollen types are presented in order of the periods of maximum abundance, but their patterns are described across all periods. Maps for present-day pollen abundances (0 ka) show the vegetation zone most closely associated with each pollen taxa, but only for pollen taxa that correspond to species that predominate in certain vegetation zones. Maps are presented in Figs. 4.12, 4.13, 4.14, 4.15, 4.16, 4.17, 4.18, 4.19, 4.20, 4.21, 4.22.

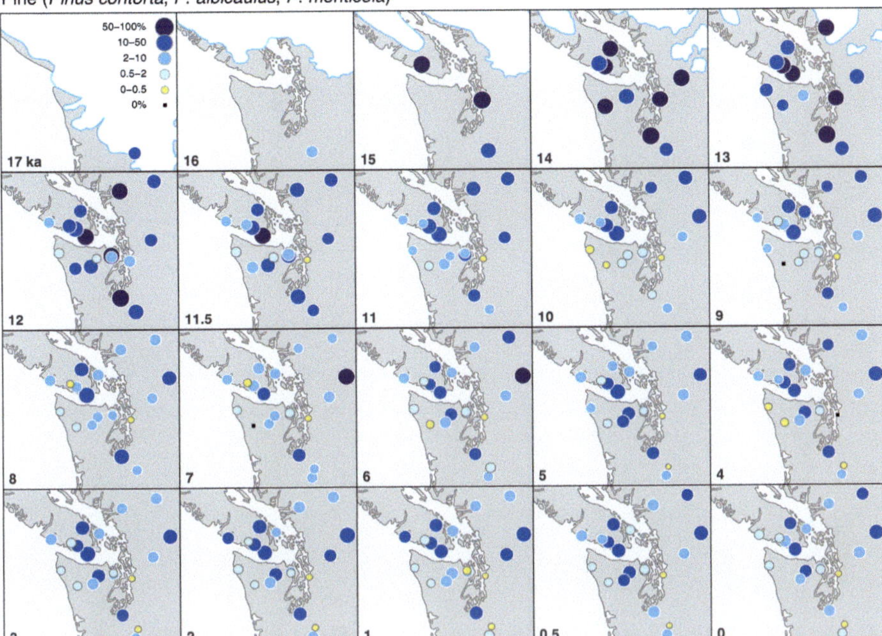

Fig. 4.12 Pine pollen percentages mapped at 20 time slices from 17 ka to present. Pine species (lodgepole pine, shore pine, whitebark pine, and western white pine) are currently widely distributed in the Pacific Northwest and as a group do not predominate in any vegetation zone. Ice extent and paleogeography are from Dyke et al. (2003). Ice extent is conservative, and does not show advances in southwest British Columbia between 14 and 13 ka. Site locations are shown in Fig. 4.11 and Table 4.2

4.5.1   Species Dominant During the Late-Glacial (14–11.6 ka): Pine, Spruce, Mountain Hemlock, and Poplar

During the late-glacial period, pine pollen is largely comprised of lodgepole/shore pine, subspecies of *Pinus contorta*. The very high abundance of pine at sites directly adjacent to the receding ice sheet suggests that this species rapidly colonized glacial outwash and/or till soils. However, it declined at 11.5 ka when the climate warmed and when a diversity of competing species able to grow to larger sizes than lodgepole/shore pine became common. Lodgepole pine likely remained abundant through the early Holocene on rocky outcrops and south facing slopes, as seen at Panther Potholes in the North Cascades (Prichard et al. 2009). The generally low levels of pine pollen throughout the Holocene in western Washington is likely a mix of lodgepole pollen on dry sites, shore pine on poor soils of the coastal plain, and western white pine in montane sites.

Spruce (*Picea sitchensis* and *P. engelmannii*)

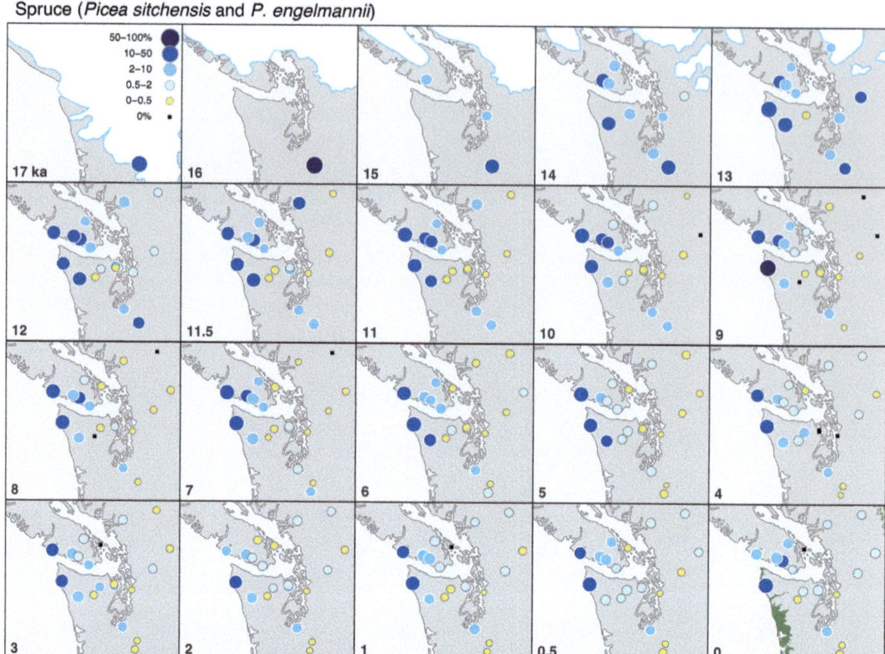

Fig. 4.13 Spruce pollen percentages mapped at 20 time slices from 17 ka to present. The map for present day (0 ka) shows the vegetaton zone in which Sitka spruce predominates. See Fig. 4.12 for details.

Spruce pollen is also abundant during the late-glacial. Although spruce pollen is indistinguishable at the species level, macrofossil evidence suggests that this pollen type mainly represents Sitka spruce with some unknown level of Engelmann spruce. Macrofossils of Sitka spruce at Yahoo Lake suggest that on the coast, Sitka spruce dominated through the late-glacial, and remained abundant until present. As Sitka spruce occurs today near periglacial environments of southeast and south-central Alaska, it is not surprising that this species was common shortly following deglaciation along the Olympic coast. Further inland in the foothills of the Cascade Mountains, a period of high spruce at Mineral Lake from 16 to 14 ka was assumed to be Engelmann spruce, as this species is common in subalpine parkland settings similar to those assumed for the late-glacial period near Mineral Lake (Tsukada and Sugita 1982). However, further north at Crocker Lake on the northeastern Olympic Peninsula and Kirk Lake in the Cascade foothills, macrofossils at 13–12 ka indicate that Sitka spruce was present in a parkland setting (Cwynar 1987; McLachlan and Brubaker 1995). A combination of genetic and fossil evidence indicates that Sitka spruce migrated northward to Haida Gwaii as early as 13.4 ka (Lacourse et al. 2005; Holliday et al. 2010). Overall, we suspect that Sitka spruce generally had a broader distribution during the late-glacial than at present.

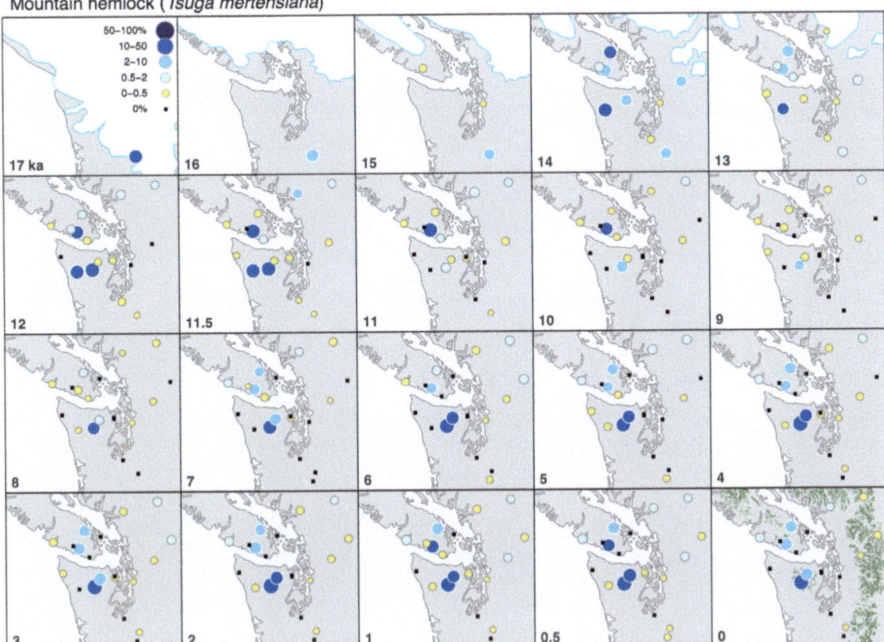

Fig. 4.14 Mountain hemlock pollen percentages mapped at 20 time slices from 17 ka to present. The map for present day (0 ka) shows the vegetation zone in which mountain hemlock predominates. See Fig. 4.12 for details.

Mountain hemlock is also a colonizer in the deglacial landscape. It increased in abundance during the increase in temperature between 15 and 14 ka on the Olympic Peninsula and on Vancouver Island. At 11.5 ka, it moved upslope to its modern subalpine elevational zone, with the exception of Martins Lake (see above) where soil development may have slowed its upslope progression. A general increase of mountain hemlock during the late Holocene is consistent with neoglacial cooling.

Poplar pollen was more common than at present during the late-glacial in the Puget Trough, especially in the foothills of the western Cascades; on the Olympic Peninsula, it also was present at lower levels (<5 % pollen). It is likely that this pollen represents the occurrence of black cottonwood in riparian forests, where glacial outwash sediments provided extensive habitat for this species. Its abundance declined into the Holocene as this habitat transitioned to less disturbed forests with organic soils.

The environmental interpretation of late-glacial conditions remains controversial because of the inherent difficulty in interpreting pollen data from this period. Most studies have interpreted the Puget Lowland as an open parkland similar to modern northern Rocky Mountain subalpine fir-Engelmann spruce forests (Barnosky 1984; Whitlock 1992). However, we note that Engelmann spruce macrofossils from the Puget Lowland generally date to older than 16 ka (e.g., at Mineral Lake and

Poplar/Cottonwood (*Populus* spp.)

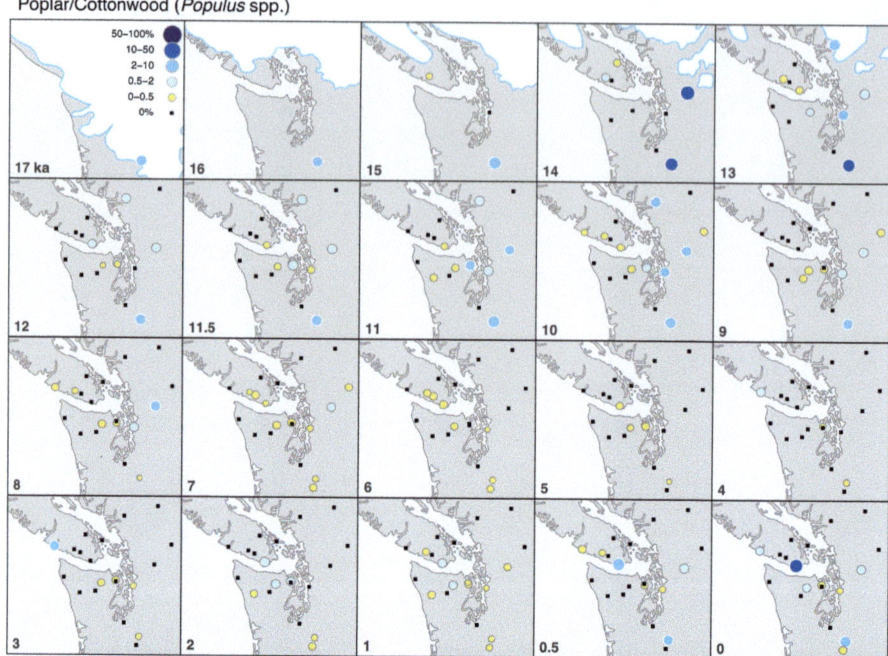

Fig. 4.15 Poplar pollen percentages mapped at 20 time slices from 17 ka to present. Poplar (cottonwood) is not abundant in this region. See Fig. 4.12 for details.

Port Moody BC; Tsukada et al. 1981; Hicock et al. 1982). After 15 ka, when the climate became distinctly more maritime, evidence of Sitka spruce, sometimes co-occurring with subalpine fir, increases (e.g., Crocker Lake at 12 ka). We suggest that the best analog for the late-glacial climatic and vegetation sequence in western Washington begins with cold maritime to subcontinental conditions that favored subalpine parkland. This transitioned at 15–14 ka to a maritime forest similar to that in coastal southeast Alaska where Sitka spruce co-occurs with mountain hemlock, cottonwood, and in a few places with subalpine fir. The presence of mountain hemlock, western hemlock, and cottonwood is more aligned with a coastal-Alaska analog than a dry-subalpine analog. The transition from continental to maritime climate remains poorly documented because most lake-sediment records begin at 14 ka. An intriguing possibility is that the two spruce species hybridized, as is possible in controlled settings (Kiss 1989). Such hybridization would contribute to the difficulty of climatic interpretation and to the difficulty in distinguishing the species in the fossil record.

To further explore the above hypothesis of full-glacial to late-glacial climates on the Olympic Peninsula, we compared paleoclimate model output for 21 ka and 14 ka with the climate data from potential analog sites. Specifically, we used paleoclimate reconstructions from the TraCE21 project (Liu et al. 2009) that were downscaled to the location of the Quillayute Airport (Forks) on the west coast of the

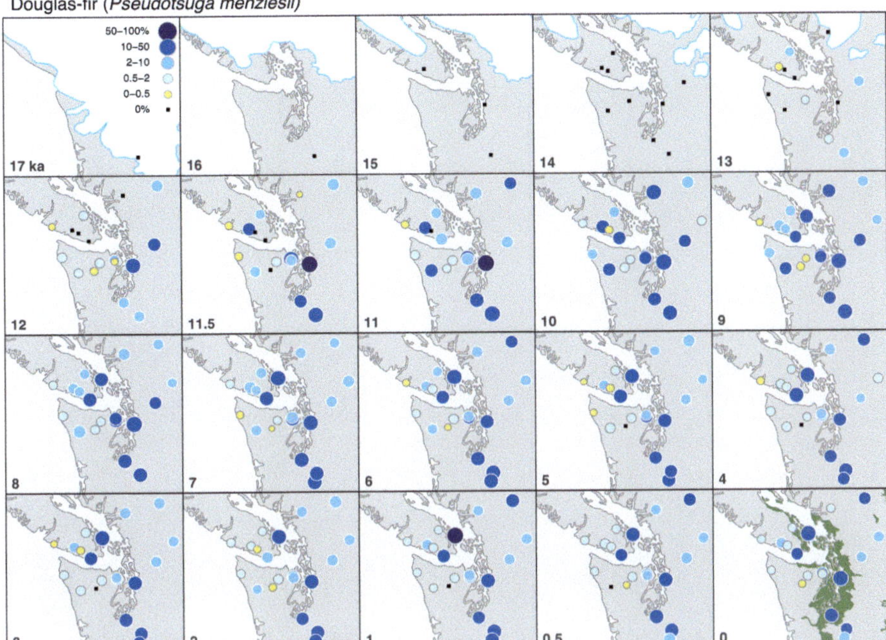

Fig. 4.16 Douglas-fir pollen percentages mapped at 20 time slices from 17 ka to present. The map for present day (0 ka) shows the vegetation zone in which Douglas-fir is most abundant, though it occurs widely across the region. See Fig. 4.12 for details.

Olympic Peninsula, following the methods in Fisher (2013). Climate diagrams were constructed for 21 ka and 14 ka. These climate diagrams were compared to three potential analog sites identified by (1) searching climate records northward along the coast until similar temperature regimes were identified and (2) selecting an interior subalpine forest to represent an Engelmann spruce–subalpine fir forest. The Bunchgrass Meadow Snotel site is in eastern Washington in subalpine Engelmann spruce–subalpine fir forest at 1520 m elevation and about 250 m below tree-line elevation. Stewart, British Columbia, is located at sea level at the head of a long fjord at the southern border of Alaska. The area supports a cool-wet western hemlock forest that grades into mountain hemlock at 600 m, and tree line is at about 1100 m. Most areas above 1400 m are covered by ice fields, which feed valley glaciers that descend to ca. 500 m. Cold Bay, Alaska, is on the Alaska Peninsula in the Bering Sea. The region supports a subarctic tundra and shrub tundra under a cold maritime climate. Glaciers extend down volcano slopes to about 700 m elevation.

On comparing the TraCE21 climate simulations with the three modern analogs, a close match can be seen between the paleo simulations and the coastal sites (Stewart and Cold Bay; Fig. 4.23). The temperature and precipitation regime of the 14-ka simulation corresponds closely with that of Stewart, in which mean winter temperatures are slightly below freezing and summer temperatures are greatly

Oak (*Quercus* cf. *garryana*)

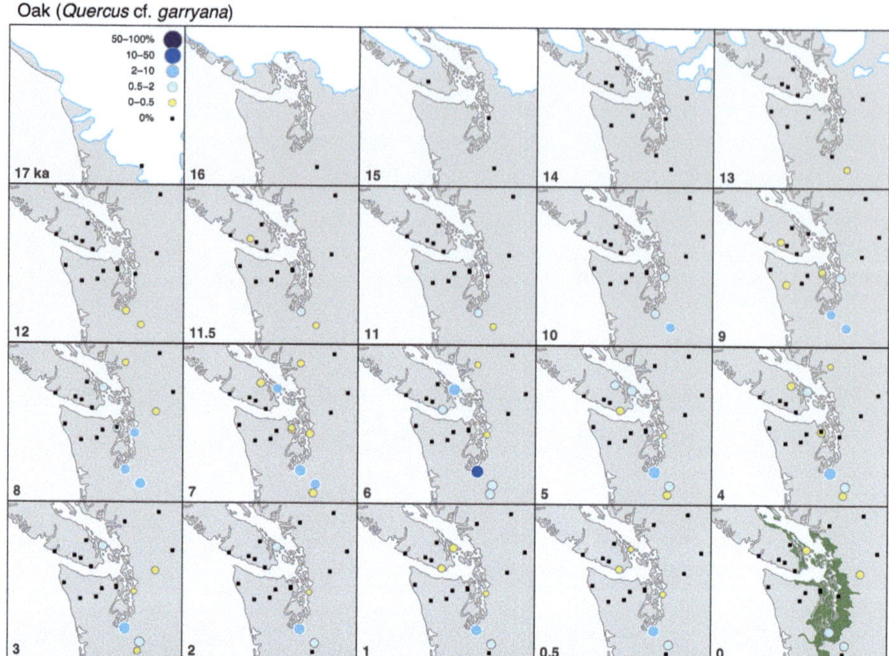

Fig. 4.17 Oak pollen percentages mapped at 20 time slices from 17 ka to present. The map for present day (0 ka) shows the vegetation zones to which oak is restricted. See Fig. 4.12 for details.

moderated. The glacial extent at Stewart is also similar to the glacial extent on the peninsula during the start of the warming at 14.7 ka (i.e., Bølling-Allerød warm interstadial event). The temperature and precipitation regime of the 21-ka simulation is similar to that of the tundra environment of Cold Bay. This cold maritime setting is far from the nearest forest as temperatures remain cool through the summer months. In contrast, Bunchgrass Meadow climate is much more continental than the simulated climate with a higher mean summer temperature and a wide diurnal temperature range. The tundra vegetation at Cold Bay is consistent with the high grass and low arboreal pollen abundance at Humptulips (Fig. 4.1), but was likely shortly followed by an interstadial event that would not be captured by the climate model. The warmer summers of these interstadials may have supported the interior spruce–fir forests observed during the Port Moody Interstadial in Vancouver (see Sect. 3.2).

4.5.2 *Species Dominant During the Early Holocene (11.6 to 6 ka): Douglas-fir, Oak, Alder, and Maple*

Douglas-fir is one of the few species that was probably absent from western Washington until the last stages of the late-glacial period. The possibility that this pollen

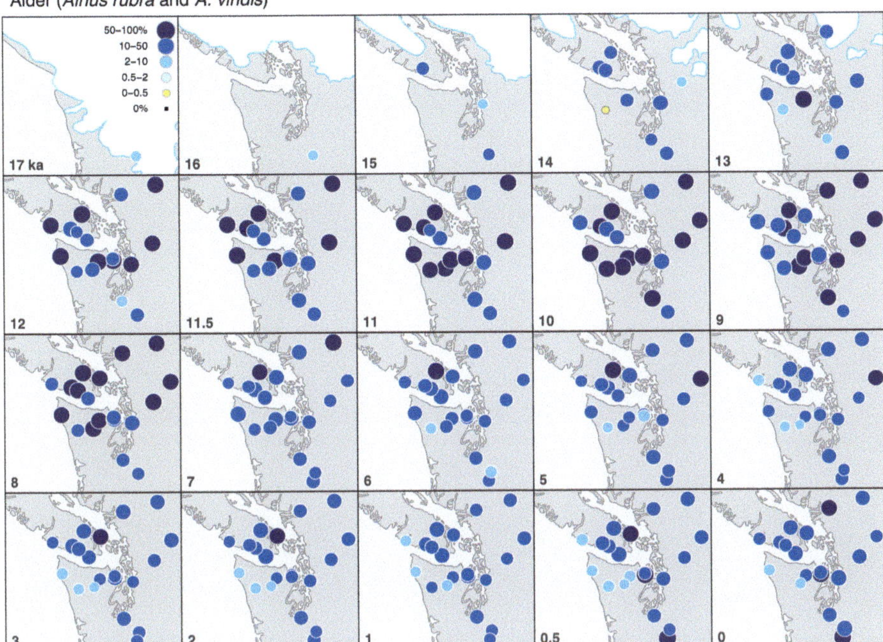

Fig. 4.18 Alder pollen percentages mapped at 20 time slices from 17 ka to present. Alder species (red alder and slide alder) are common over the entire region today. See Fig. 4.12 for details.

type represents western larch is remote, as larch occurs only east of the Cascade Range. Douglas-fir pollen is not identified at any well-dated location prior to 13.5 ka, thus contrasting with a previous synthesis of Douglas-fir that claimed an LGM refugium for Douglas-fir was possible near the Puget Lowlands (Gugger and Sugita 2010). At 13 ka, it is present at low levels ($<5\%$) in the Puget Lowland and southern Vancouver Island, though the radiocarbon chronology is poor at the sites that indicate its presence at this time. However, its pollen dramatically increased in abundance and macrofossils are found in the early Holocene (11.5 ka). As currently mapped, the pollen evidence supports the idea of a rapid northward migration at low abundance followed by a rapid increase in abundance upon early Holocene warming. Better radiocarbon chronologies may reveal a wave-like migration closer to 12 ka, perhaps with a migration lag to Vancouver Island. Douglas-fir pollen peaks in the early Holocene as it was favored in western Washington by the warmer summers and more frequent fire. Although Douglas-fir pollen declines in abundance at the most maritime sites by 8 ka, it remains fairly abundant in the Puget Lowland and eastern Vancouver Island into the late Holocene.

Oak pollen, representing Oregon white oak, the only oak species occurring in the Pacific Northwest, first appears at 13 ka in the southern Puget Lowland and becomes more common from 10 to 6 ka. As for Douglas-fir, the warm and dry climate and the frequent fires of the early Holocene likely allowed expansion of the oak

Maple (*Acer macrophyllum* and *A. circinatum*)

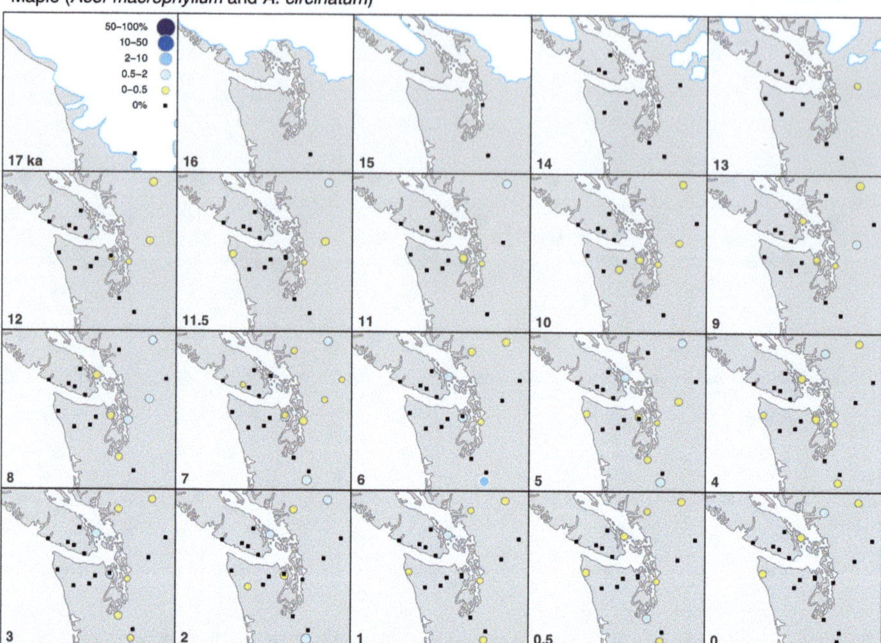

Fig. 4.19 Maple pollen percentages mapped at 20 time slices from 17 ka to present. Maple species (bigleaf maple and vine maple) are widely distributed and most common in the understory of riparian forests. See Fig. 4.12 for details.

prairies in the Puget Lowland (Leopold and Boyd 1999). Oak pollen increased later on Vancouver Island (at 8.3 ka), possibly due to a migration lag across the San Juan Islands, or due to low winter temperatures that precluded its presence until more maritime climate prevailed (Pellatt et al. 2001). Oak pollen increased at a similar time on the mainland of British Columbia at the same latitude, providing support for climatic control of its range expansion (Mathewes 1973).

Alder pollen represents two common species: red alder and slide alder. Both produce abundant pollen. Because too few studies distinguished the pollen, the mapped values show total alder percentages. The period of regionally highest abundance, from 12 to 8 ka, coincides with the period of high Douglas-fir. As red alder is a major nitrogen fixer, its abundance during the early Holocene likely greatly enhanced nitrogen availability (Compton et al. 2003).

Maple is poorly represented in the pollen record because maple species are primarily insect pollinated and thus produce much less pollen than wind pollinated conifer and broadleaved species. Big-leaf maple is likely the largest contributor of maple pollen. Although maple pollen almost never exceeds 2 %, it is more common in the middle Holocene. There is little spatial patterning in its pollen abundance.

The interpretation of early Holocene vegetation and disturbance regimes from these mapped data is similar to that discussed above for the Olympic Peninsula

True firs (*Abies amabilis, A. lasiocarpa*, and *A. grandis*)

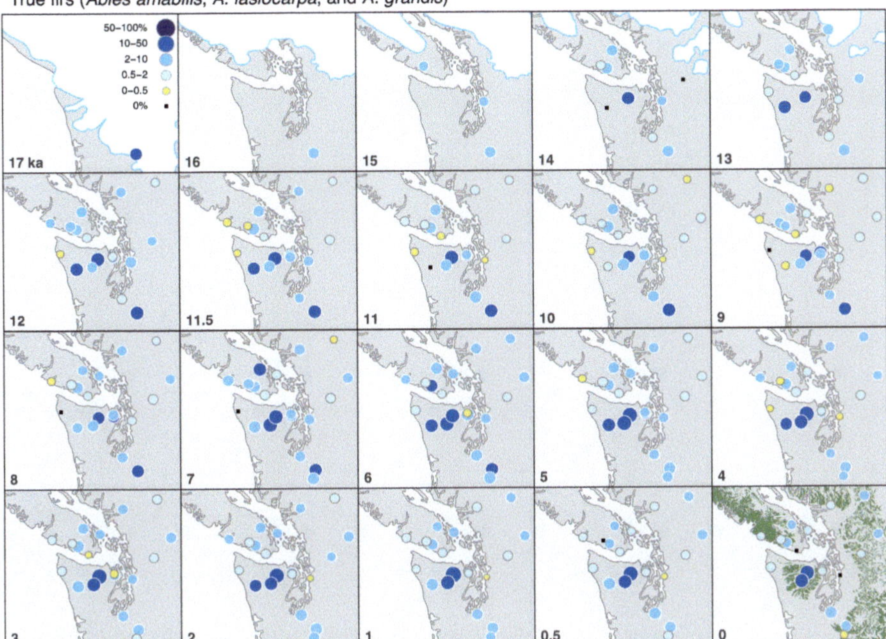

Fig. 4.20 True fir pollen percentages for 11 major pollen taxa mapped at 20 time slices from 17 ka to present. The map for present day (0 ka) shows the vegetation zones in which true fir taxa (Pacific silver fir, subalpine fir, and grand fir) predominate. See Fig. 4.12 for details.

sites, however, there remains complexity in the interpretation of the early Holocene forest dynamics. The warm and dry summers implied by the high fire frequency is consistent with an increase in oak and Douglas-fir, but not with an increase in red alder. Red alder today on the western Olympic Peninsula is restricted to disturbed riparian forests, but can extend upslope in wet areas following disturbance. The very high red alder abundance, however, does not appear consistent with the inferred drier conditions of the early Holocene.

An interpretation of the early Holocene Douglas-fir–red alder association may lie by simply summoning a modern analog in the central Oregon Coast Range. Mean July temperatures in the Oregon Coast Range are at least 4 °C warmer than the western Olympics, similar the temperature anomaly at that time (Fig. 3.4). The historical fire regime in the Oregon Coast Range was marked by very large infrequent fire which ranged from mixed severity to high severity fire (Impara 1997), possibly quite similar to the rate of burning inferred on the western Olympic Peninsula. The presettlement pollen assemblage at Little Lake in the Oregon Coast Range has proportions of red alder and Douglas-fir similar to that at Yahoo Lake in the early Holocene (Long et al. 2007). Studies of ecological succession in the Oregon Coast Range reveal how red alder is a more serious competitor with Douglas-fir on hillsides after fire than it is on the Olympic Peninsula. The Plant Association Guide

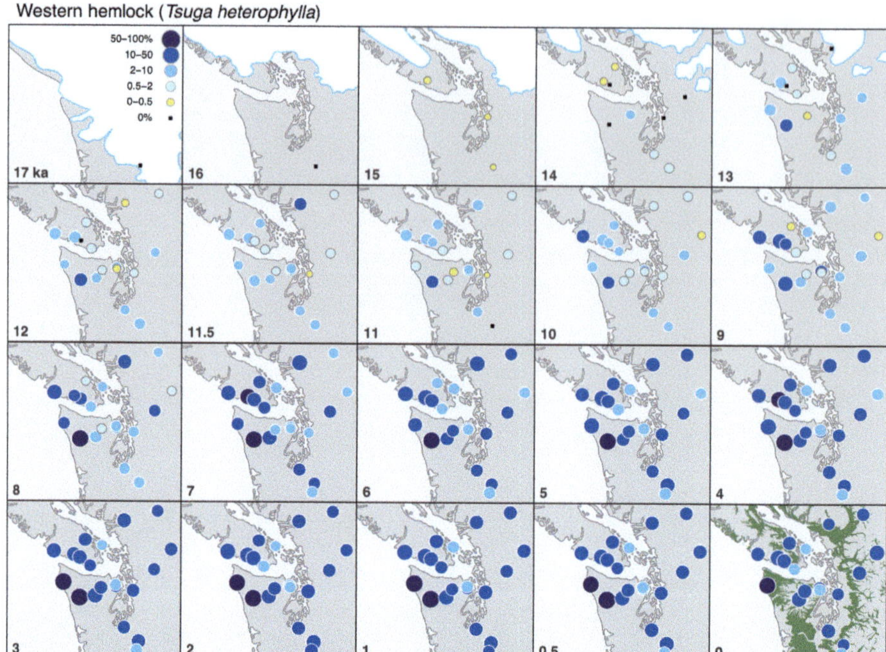

Fig. 4.21 Western hemlock pollen percentages mapped at 20 time slices from 17 ka to present. The map for present day (0 ka) shows the vegetation zones in which western hemlock predominates, though it occurs throughout the region. See Fig. 4.12 for details.

for the Siuslaw National Forest in west-central Oregon states that many sites undergo a red-alder-dominated stage following fire, that red alder germinates readily on exposed mineral soil in full sunlight, and it overtakes and suppresses conifers after three years (Hemstrom and Logan 1986). If conifer seed sources are limiting, then after 100 years the alders senesce and the site reverts to a brushfield with scattered large Douglas-fir. This competition is so widespread on the steep slopes of the Coast Range that broad-scale herbicide application is common to prevent alder from overtopping the more valuable Douglas-fir after harvest.

The spatial gradient of Douglas-fir pollen from the western slope to the Puget Lowland indicates that fire and lower precipitation in the Puget Lowland allowed for a great expansion of Douglas-fir forests. However, a similar spatial gradient was not observed for alder. It is likely that alder was very aggressive colonizing sites in the wetter climate of the western Olympics, but in the Puget Lowlands fire may have been more common providing frequent opportunities for alder establishment. We speculate that alder was more competitive in the west balanced its more frequent opportunity for establishment in the east.

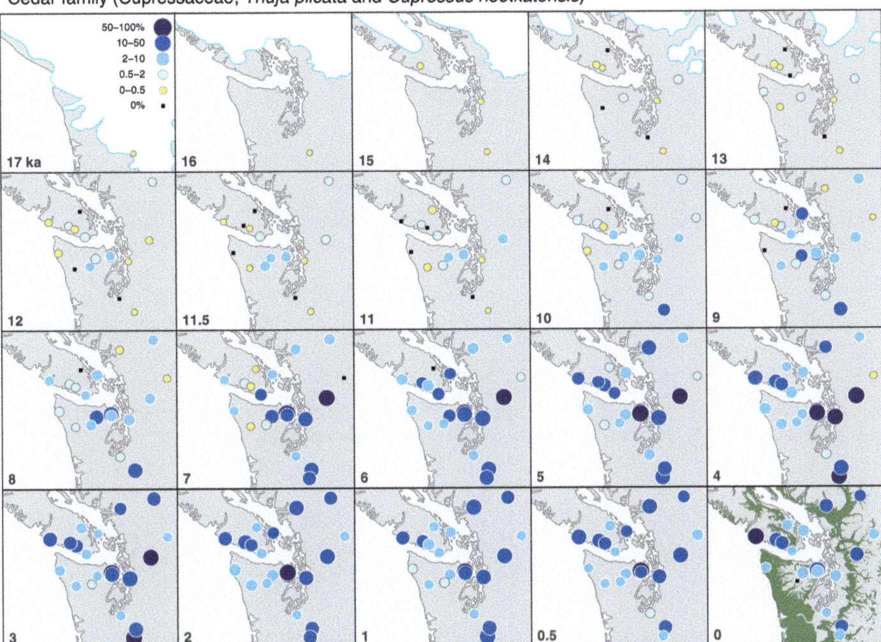

Fig. 4.22 Cedar-family pollen percentages mapped at 20 time slices from 17 ka to present. The map for present day (0 ka) shows the vegetation zones in which cedar species (western red cedar and Alaska yellow cedar) predominate. This pollen type also reflects juniper, which is locally common in dry subalpine meadows. See Fig. 4.12 for details.

4.5.3   Species Dominant During the Middle to Late Holocene (6 ka to Present): True Firs, Western Hemlock, and Western Redcedar

True-fir pollen is represented by two common species in the region: subalpine fir and Pacific silver fir. As these species cannot be distinguished by pollen and occur in quite different climatic zones, it is not surprising that there is no strong spatial patterns in the true-fir pollen data. However, because fir pollen is large and poorly dispersed, the presence of fir pollen can aid interpretations because its presence more likely reflects local presence than does the pollen of other tree species. Macrofossils, which can be identified to species, show that the late-glacial fir was likely subalpine fir in the Puget Lowland and Pacific silver fir on the west coast (Cwynar 1987; Gavin et al. 2013). The early Holocene climate did not favor Pacific silver fir even at the wettest sites, while rainshadowed high-elevation sites were favorable to subalpine fir. The late Holocene increase in true-fir pollen represents an expansion of Pacific silver fir in late-successional forests on the western Olympic Peninsula.

Table 4.2 Late Quaternary pollen records and other sites mentioned in the text

Site	Name	Latitude	Longitude	Elev. (m)	Lake size (ha)	References
1	Frozen Lake	49° 36′N	121° 28′W	1180	3	Hallett et al. 2003
2	Pinecrest Lake[a]	49° 30′N	121° 26′W	320	1.8	Mathewes et al. 1972; Mathewes and Rouse 1975
3	Squeah Lake	49° 29′N	121° 24′W	250	2.2	Mathewes and Rouse 1975
4	Marion Lake[a]	49° 19′N	122° 33′W	305	16	Mathewes 1973; Mathewes and Heusser 1981
5	Mike Lake[a]	49° 16′N	122° 32′W	225	4.8	Pellatt et al. 2002
6	Mt. Barr Cirque	49° 16′N	121° 31′W	1376	2	Hallett et al. 2003
7	Clayoquot Lake	49° 12′N	125° 30′W	17	47	Gavin et al. 2003
8	Porphyry Lake[a]	48° 54′N	123° 50′W	1100	0.3	Brown and Hebda 2003
9	Roe Lake[a]	48° 47′N	123° 18′W	100	3.0	Lucas and Lacourse 2013
10	Mosquito Lake Bog	48° 46′N	122° 07′W	198	–	Hansen and Easterbrook 1974
11	Whyac Lake[a]	48° 40′N	124° 51′W	10	2.5	Brown and Hebda 2002, 2003
12	Panther Potholes[a]	48° 40′N	121° 02′W	1100	0.3	Prichard et al. 2009
13	Mt Constitution	48° 39′N	122° 50′W	660	–	Sugimura et al. 2008
14	Saanich Inlet	48° 38′N	123° 30′W	0	–	Pellatt et al. 2001
15	Ayer Pond	48° 37′N	122° 49′W	50	–	Kenady et al. 2011
16	Pixie Lake[a]	48° 36′N	124° 12′W	70	5.3	Brown and Hebda 2002, 2003
17	Walker Lake[a]	48° 32′N	124° 00′W	950	0.4	Brown and Hebda 2003
18	East Sooke Fen[a]	48° 21′N	123° 41′W	155	0.1	Brown and Hebda 2002
19	Kirk Lake[a]	48° 15′N	121° 37′W	190	0.6	Cwynar 1987
20	Ebey's Prairie	48° 12′N	122° 42′W	25	–	Weiser and Lepofsky 2009
21	Ahlstrom's Prairie	48° 10′N	124° 43′W	60	–	Anderson 2009
22	Manis Mastadon Site	48° 03′N	123° 07′W	170	–	Petersen et al. 1983; Waters et al. 2011
23	Wentworth Lake[a]	48° 01′N	124° 32′W	47	14	This study
24	Crocker Lake[a]	47° 56′N	122° 53′W	54	26	McLachlan and Brubaker 1995
25	Cedar Swamp[a]	47° 54′N	122° 53′W	57	–	McLachlan and Brubaker 1995
26	Moose Lake[a]	47° 53′N	123° 21′W	1544	3.7	Brubaker and McLachlan 1996; Gavin et al. 2001
27	Royal Basin	47° 49′N	123° 13′W	1770	–	Gavin and Brubaker 1999

Table 4.2 (continued)

Site	Name	Latitude	Longitude	Elev. (m)	Lake size (ha)	References
28	Hall Lake[a]	47° 48′N	122° 19′W	104	3.1	Sugita and Tsukada 1982
29	Martins Lake[a]	47° 43′N	123° 32′W	1423	0.8	Gavin et al. 2001
30	Yahoo Lake[a]	47° 41′N	124° 01′W	717	3.7	Gavin et al. 2013
31	Queets Hollows	47° 38′N	123° 58′W	120	–	Greenwald and Brubaker 2001
32	Humptulips Bog	47° 17′N	123° 55′W	120	–	Heusser et al. 1999
33	Nisqually Lake[a]	47° 02′N	122° 38′W	65	8	Hibbert 1979
34	Reflection Pond 1	46° 46′N	121° 43′W	1482	0.6	Dunwiddie 1986
35	Mineral Lake[a]	46° 44′N	122° 10′W	436	103	Sugita and Tsukada 1982
36	Davis Lake[a]	46° 32′N	122° 15′W	290	–	Barnosky 1981

Site numbers refer to Fig. 4.11

[a] Indicates sites used for mapping pollen abundances

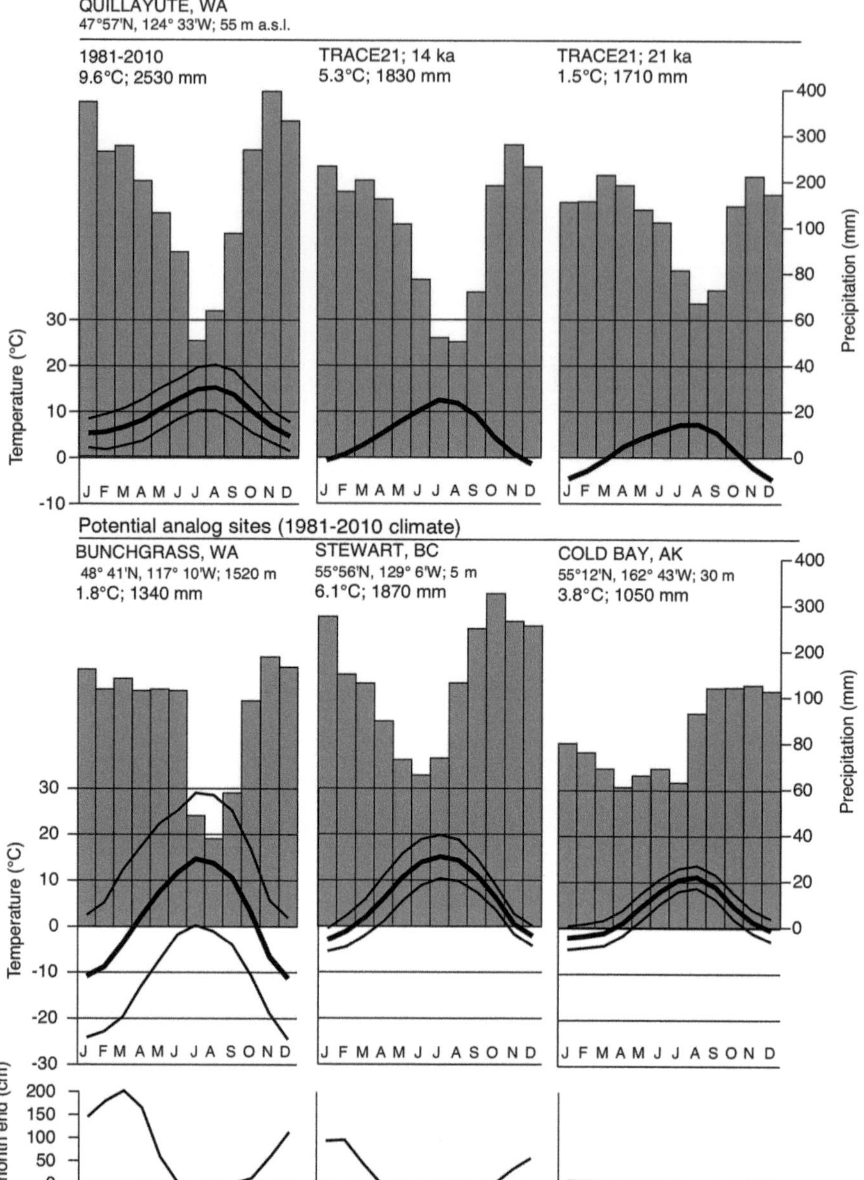

Fig. 4.23 Climate analogs for the LGM (21 ka) and the late-glacial (14 ka) on the Olympic Peninsula coast. The top row shows climate diagrams for Quillayute (near Forks) and climate simulations from the TraCE21 simulation (Liu et al. 2009) downscaled to the same location for 14 ka and 21 ka following the methods in Fisher (2013). The bottom row shows climate diagrams and monthly snow depth for three potential analog sites for LGM and late-glacial climates. Site coordinates, elevation, and mean annual temperature and precipitation are shown for each site

Western hemlock, the most common late-successional tree species below the subalpine elevations in western Washington, follows a similar history to Pacific silver fir, but because its pollen is so widespread its spatial pattern on the pollen maps is less distinct. Western hemlock is present by 14 ka mainly in the Puget Lowland and Vancouver Island. Just before 13 ka, it increases rapidly on the western Olympics but then declines to low levels (<5%) at 11.5 ka. It remains most abundant on the Pacific slope through the early Holocene, and increases to more than 10% in the Puget Lowland by 7 ka.

The cedar family pollen represents two tree species (western redcedar at low elevations and Alaska yellow cedar in the subalpine) and possibly scattered junipers (common juniper and seaside juniper). These species are, at most sites, poorly represented by pollen. It is particularly difficult to determine which cedar species was important on the landscape during the long period of low pollen abundance from 15 to 11 ka. Macrofossils, which can be identified to species, indicate that increases in cedar pollen at Moose Lake at 9 ka is likely entirely Alaska yellow cedar. In contrast, macrofossils at low elevation indicate that western redcedar was the predominant cedar type and indicate a clear northward progression of cedar pollen, from 10 ka at Mineral Lake to 6 or 4 ka on Vancouver Island. Northward spread from a southern population is supported by genetic data that suggest it occurred in a southern refugium south of Washington during the LGM (O'Connell et al. 2008). The importance of this northward progression for Native American prehistory was noted by Hebda and Mathewes (1984), as large cedar logs that are keystone of coastal cultures were likely not available until the large increase in cedar pollen.

The late-Holocene-mapped pollen data are consistent with a regionally uniform increase in available moisture and a decrease in fire that promoted the increase in shade-tolerant tree species. While climatic fluctuations still occurred, they may have been of lower magnitude than those of the the early Holocene. Additionally, forests dominated by long-lived trees species may have experienced significant inertia to climate change. Paleoecological sites that record a more local signal of vegetation, such as macrofossil records from small ponds, are more likely to reveal coherent fluctuations in species abundances than can be seen in the time-slice maps of pollen data (Dunwiddie 1986).

4.5.4   Alternative Approaches to Interpreting Postglacial Vegetation Change

Although the sequence of postglacial vegetation change varies greatly among the five sites presented above, the mapped data reveal some major common patterns. In particular, over the entire region the tree taxa reach peak abundances in coherent patterns, strongly indicating a regional climatic forcing and relatively little influence of migrational lags. Very broadly, the late-glacial favored species adapted to cool summers, the early Holocene favored species adapted to fire disturbance during warmer summers, and the late Holocene favored late-successional species under a less-frequent disturbance regime. Although the magnitude of vegetation change varied among sites, and detailed sites provided evidence of shorter-term fluctuations, each pollen record contains elements of this sequence. As is the common practice in paleoecology, the vegetation changes at a particular site have been explained in a qualitative approach using species traits (e.g., the inferred climatic thresholds and life history strategies) and the climatic and disturbance history. Newer methods applied to the Pacific Northwest postglacial pollen records deserve a mention for their potential to improve our understanding of the proximal causes of vegetation change.

Statistical methods developed for functional community ecology have recently been applied to data from pollen records. These methods link species quantitative traits (e.g., shade tolerance and longevity) to changing environmental conditions to provide a functional explanation for observed vegetation changes. Lacourse (2009) first applied this method to a pollen record from Misty Lake on Vancouver Island, a pollen record similar to that from Yahoo Lake. Two methods were used. One method, RLQ analysis, is a multivariate ordination of species abundance data that are constrained by those species traits and by environmental data (Dolédec et al. 1996). The other method, fourth-corner analysis, tests for correlations between species trait data and the environmental data (Legendre et al. 1997). Lacourse (2009) found significant correlations between the traits of the woody plant species and the temporal changes in environmental factors. The cold wet climate of the late-glacial favored species adapted to open environments that are fast growing, short lived, and shade intolerant such as slide alder and lodgepole pine. The warm-dry and high-insolation favored species that reach greater heights such as Sitka spruce (and Douglas-fir at Yahoo Lake). The more stable, cool-wet late Holocene favored large-seeded, slow-growing, and shade-tolerant species such as Pacific silver fir and western hemlock. The results also pointed to a suite of plant traits that are significantly correlated to environmental variables.

Dynamic global vegetation models (DGVMs) simulate changing abundances of broad categories of species (PFTs) responding to a changing environment, and include explicit modeling of plant physiology and soil water abundances. Parameterizing the plant functional types (PFTs) to represent particular species is challenging. Few have attempted to use a DGVM to attempt to re-create the observed changes in a pollen diagram. Following the example by Miller et al. (2008), Fisher (2013) used the Lund-Potsdam-Jena General Ecosystem Simulator (LPJ-GUESS) model to attempt to simulate the vegetation changes at the sites of the five pollen records on the Olympic Peninsula. The model input included a 14,000-year climate simulation that was downscaled to each of the five sites. In parameterizing the PFTs to the species level required for the model, it was apparent that persisting snow pack was the predominant driving factor of the biogeography of species relative abundances across the peninsula. Fisher (2013) modified LPJ-GUESS to restrict the growing season to the snow-free period. This adjustment resulted in both spatial patterns in species abundance, and temporal patterns at each site, to more closely follow the observed changes. These results suggest that the impact of changing climate on vegetation will be enhanced by the large amount of "at risk" snowpack that exists on the peninsula (Nolin and Daly 2006).

Chapter 5
Brief Review of the Archeological Record in a Context of Environmental Change

Abstract The detail emerging from the paleoenvironmental record provides important context for the archeological record. Some of the oldest evidence of humans in the Americas is found in the Pacific Northwest. In this very brief chapter, we examine the correlation of dates of important archeological sites with the well-dated events of the paleoenvironmental record described in Chap. 4. We give special consideration to relative sea-level history, and to the paleoseismic history with respect to the archeological record.

5.1 Introduction

Humans have inhabited the Olympic Peninsula for most of the postglacial period and they experienced a dynamic environment throughout that time. While the complex and incompletely understood sea-level history presents a challenge for archeologists, several robust and well-dated discoveries in the area provide an opportunity to correlate evidence of human prehistory with the paleoecological record. Here, we present a brief review of some well-studied sites and discuss their paleoenvironmental context. Our goal is to place the dates of known sites within the context of paleoecological records, not to review cultural inferences obtained from these sites. Early reviews of Olympic Peninsula prehistory, and the difficulty of obtaining well-dated sites, are presented in Bergland (1983) and Schalk (1988). A review of Northwest Coast archeology is presented in Moss (2011). A more general introduction to San Juan Island Coast Salish archeology is in Stein (2000).

5.2 The Manis Mastodon Site (Late-Glacial Period)

This site, south of the town of Sequim in the rain shadow of the Olympic Mountains, was discovered in 1977 during excavation for a pond in a pasture. The discovery of this mastodon, and associated bison, caribou, and muskrat, is recounted in several places (Gustafson et al. 1979; Kirk and Daugherty 2007). The significance of the find to human history was immediately recognized with discovery of a possible

© Springer International Publishing Switzerland 2015
D. G. Gavin, L. B. Brubaker, *Late Pleistocene and Holocene Environmental Change on the Olympic Peninsula, Washington*, Ecological Studies 222,
DOI 10.1007/978-3-319-11014-1_5

bone projectile point embedded in the mastodon rib. The age of the mastodon was determined by radiocarbon dates of wood associated with the bones, with the most precise date yielding a calibrated age of 13.6–13.8 ka (Gustafson 1979). This date as a human-associated artifact is of great significance, as earliest North Americans were long thought to be from the Clovis period occurring briefly from ca. 13.1 to 12.8 ka and marked by distinct projectile points and tools (Waters and Stafford 2007). More recently, Waters et al. (2011) obtained additional evidence on the embedded bone. A CT scan indicated that the embedded bone was a sharpened mastodon bone and bone growth indicated that the mastodon was not killed directly by the injury. Accelerator mass spectrometry (AMS) radiocarbon dates on bone collagen place the age at 13.76–13.86 ka. This age is remarkably similar to bison bones with possible butchering marks found in Ayer Pond on Orcas Island, which had a bone collagen radiocarbon age that calibrated to 13.78–13.89 ka (Kenady et al. 2011). The statistically identical age of these two sites, both discovered accidentally, suggests this was an important time of human presence in the region. Both sites, in shallow water, are consistent with a practice of storing carcasses in bogs that would freeze over (Kirk and Daugherty 2007). Despite these patterns and growing evidence for pre-Clovis people in the Pacific Northwest (Gilbert et al. 2008; Jenkins et al. 2012a), both of these pre-Clovis-aged sites could benefit from additional evidence of human presence.

The date of these discoveries at 13.8 ka is within a period of great climatic variability. The pollen record from the peat soils at the mastodon site revealed open meadow conditions persisted from 13.8 to 12.9 ka (Petersen et al. 1983). The presence of prickly pear cactus pollen and hornwort seeds, both species that have northern limits close to the Olympic Peninsula today, suggests that summer temperatures were similar to today. The open character of the site during the late-glacial suggests that it was drier than present, as Douglas-fir forests occupied the site before recent land clearance. In contrast to the evidence of open vegetation at Manis, our two best-dated pollen records on the peninsula (Yahoo and Wentworth lakes on the western peninsula) and Pixie Lake on Vancouver Island (Brown and Hebda 2002) show that 13.8 ka dates to within 100 years of a rapid transition to dense forests. At these sites, pine pollen decreased and the pollen of mesic conifer species increased, including Sitka spruce, Pacific silver fir, mountain hemlock, as well as the initial increase of western hemlock. That these vegetation changes were synchronous at three well-dated sites and that multiple species increased in abundance suggest a large increase in moisture and possibly an increase in winter temperature (which may have previously precluded cold-sensitive species such as western hemlock). The pollen record at Manis, however, did not show increased forest density cover until ca. 12.9 ka (Petersen et al. 1983). A more significant increase in forest density at Manis occurred at 10 ka when western hemlock increased and pollen of open-site taxa such as buffaloberry (*Shepherdia*) and willow decreased[1]; Wentworth Lake

---

[1] The lack of Douglas-fir pollen at Manis is anomalous with respect to the nearby pollen record from Crocker Lake (Fig. 4.8).

also shows increased western hemlock pollen at this time. It is likely that the increase in moisture at 13.8 ka was greatly muted within the rain shadow.

The large increase in moisture at 13.8 ka would have steepened the precipitation gradient on the peninsula and thus would have increased the climatic uniqueness (and probably the relative habitability for people) of the rain shadow environment. The increase in moisture at this time is also consistent with evidence for glacial advances in the Pacific Northwest. The well-studied Sumas moraines on the lower mainland of British Columbia date to 13.7–13.2 ka, indicating valley glaciers descended the Fraser River close to the modern coastline (see Sect. 4.3). The northeastern Olympic Peninsula would have been an oasis between a periglacial environment on the mainland and cool and wet climate on the west coast. Open habitat and forest edge conditions, and steep environmental gradients, would have provided a range of habitat types within short travel distances. Furthermore, lower sea level at that time (Fig. 5.1) may have resulted in greater connectivity across the San Juan Islands and resulting in a "filter bridge" for colonization of Vancouver Island by several species of megafauna (Harington 1975; Wilson et al. 2009). After this period of general increased forest density on the peninsula (except at Manis), the colder and likely drier climate of the Younger Dryas (12.9–11.6 ka) is marked by very little archeological and faunal evidence in the Pacific Northwest (Fedje et al. 2011).

5.3   Late Holocene Cultural Phases

The scarcity of archeological sites dating to the early Holocene, and the few sites dating to the late Holocene, precludes making strong correlations between environmental changes and cultures on the Olympic Peninsula. Most dates of early Holocene artifacts are spear tips and tools on mountain ridges that indicate hunting and gathering activities (McNulty 2009). In contrast, coastal sites date to either the late-glacial during a time of higher-than-modern sea level (e.g., Ayer Pond, where the bison dates to a time the site was near sea level), or to the late Holocene (after 4 ka). A recently published sea-level curve for southeastern Vancouver Island (James et al. 2009) suggests that the lack of Younger-Dryas- and early-Holocene-aged coastal sites is due to sea-level rise that likely submerged archeological sites. After 4 ka, sea level rose only gradually to present (ca. 1 m in the Gulf Islands of Vancouver Island), and indeed, coastal archeological sites date to within 4 ka (Fedje et al. 2009; Fig. 5.1). Attrition of archeological sites also occurs by coastal erosion (Moss 2011). For the late Holocene, however, there is evidence of changing culture and resource use that can be placed into the context of the new paleoenvironmental reconstructions presented here. For a review in a larger geographic context, see Moss (2011).

The Locarno Beach Cultural Phase of the Straits of Georgia region occurred from 3.5 to 2.4 ka (Croes 1989). The archeological record indicates that this is a period of increased storage economy (Croes 1989). Climate and vegetation records indicate that Locarno Beach was a period of increasing cool and wet climate. For

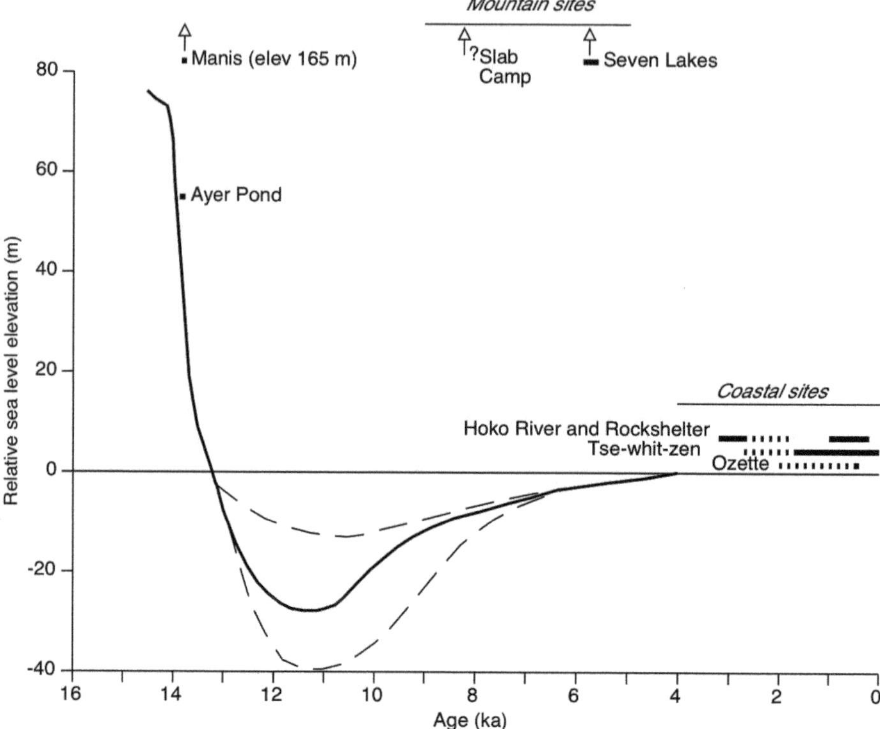

Fig. 5.1 Sea-level curve for southeastern Vancouver Island showing the combined effects of isostasy and sea-level rise on relative sea level (James et al. 2009). The *dashed lines* represent the range of possible values constrained by radiocarbon dates, and the *solid line* is the preferred curve by James. Note a sea-level rise of almost 10 m between 8 and 4 ka. Dates of important archeological sites are plotted above the sea level curve. *Solid lines* indicate age ranges of occupancy of the sites and *dashed lines* connect outlier dates to the remainder of dates at the site. Note that the coastal site elevations are approximate. See Schalk (1988) for a full list of dates

example, the Tiedemann Advance in the southern Coast Mountains occurred from 3.3 to 2.9 ka (Fig. 3.4, Menounos et al. 2009). At this time, the detailed record at Yahoo Lake shows an absence of fire and an increase in late-successional Pacific silver fir and western redcedar. The increase in redcedar is notable in that large cedar trees, and thus the woodworking technologies that depend on this resource, were likely not common until the middle to late Holocene (Hebda and Mathewes 1984). Indeed, artifacts constructed from large redcedar date to within the last 5000 years (Bergland 1983; Hebda and Mathewes 1984). Interestingly, sea-surface temperatures increase at 3.3 ka, which could have resulted in increased moisture advecting onshore (Barron et al. 2003). The only well-dated site on the Olympic Peninsula in this period is the Hoko River site on the northern Olympic coast, which spans the period from 3.2 to 2.7 ka (calibrated ages presented in Ames 2005). The site was interpreted as a fishing camp, with hook and net artifacts, and provided evidence of the development of both marine and riverine fishing (Croes and Hackenberger 1988). There may have been a general shift from hunting land mammals to increased fishing and use of coastal resources during the middle Holocene, associated with stabilization of sea levels and/or increased streamflow in the wetter climate of the late Holocene. However, this interpretation may be confounded by sea-level rise that censored early Holocene coastal sites. A review of zooarcheology of the Northwest Coast does reveal increasing proportion of salmonids in middens through the late Holocene but also broad-spectrum foraging was common throughout the Holocene (Butler and Campbell 2004).

The Marpole Cultural Phase (2.4–1.2 ka), of the Fraser River valley and the Gulf Islands (Lepofsky et al. 2005), is marked by increased numbers of large houses and settlements of multiple houses, regional art forms, and symbols of prestige (Grier 2003). While there are fewer sites dating to this general period on the Olympic Peninsula than on the Georgia Strait and Fraser River, the recent excavation of the large village of Tse-whit-zen in Port Angeles is an exception. This site was occupied from 1.7 ka to historic times, though the oldest date suggests occupation at 2.7 ka (Mapes 2009). This is a time of an absence of glacial advances and an increase in fire occurrence throughout the Pacific Northwest (Gavin et al. 2007), including subalpine meadows in southwest British Columbia (Hallett et al. 2003), in lowland prairies on Whidbey Island east of the Olympic Peninsula (Weiser and Lepofsky 2009) and at Ozette Prairie on the west coast of the Olympic Peninsula (Gutchwesky 2004; Anderson 2009). Fire increased sharply at 2.4 ka at Yahoo Lake (Fig. 4.5) and ca. 2 ka at Martins Lake (Fig. 4.7), the latter located in subalpine meadows. Lepofsky et al. (2005) discussed the cultural and environmental changes at the time. Warmer and drier conditions may have made the Fraser Valley more desirable because such a climate would have promoted a variety of rich ecosystems and increased availability of a range of food resources (Lepofsky et al. 2005).

The Ozette village site deserves special mention because of the excellent preservation and large number of artifacts (Kirk and Daugherty 2007). This site is a set of at least four cedar plank houses that were buried in a mudflow dating to within the last 500–300 years, though the site was occupied before and after the burial. The quick burial and water saturation resulted in excellent preservation of over 55,000 wood and fiber artifacts (Ames 2005). The dating of the Ozette site is problematic because of the lack of suitable material for accurate aging of the mudflow. Hutchinson and McMillan (1997) interpret the dates to indicate occupation at least 800 years ago, and dendrochronological work on cedar planks suggests the mudslide occurred after AD 1718. More recently, the initial estimated age of the mudslide of ca. AD 1500 has been reevaluated in light of paleoseismic evidence and it has been suggested the great AD 1700 earthquake may have triggered the mudslide (Kirk and Daugherty 2007). Even more recent paleoseismology studies from marine turbidites have shown that an earthquake dating to AD 1500 occurred along the Olympic Peninsula (the "T2" event in Goldfinger et al. 2012).

The seismic history of the Cascadia Subduction Zone suggests coastal villages were greatly affected by earthquakes and tsunamis. Some paleoseismic studies suggest that subduction zone earthquakes recur on a roughly 500-year interval, with only the most recent event in AD 1700 (Atwater 1987; Atwater et al. 2005; Nelson et al. 2006). Hutchinson and McMillan (1997) found that hiatuses in the radiocarbon dates of cultural artifacts correlate strongly with paleoseismic and tsunami events. This correlation is strong at sites where coastal geomorphology would amplify tsunamis, such as in Nootka Sound on Vancouver Island, but the correlation is less strong on the straight Olympic coast. A strong oral tradition recounting effects of earthquakes and tsunamis occurs among the Nuu-chah-nulth of western Vancouver Island (McMillan and Hutchinson 2002). The Makah of the northwestern Olympic Peninsula also have an oral tradition of a catastrophic flood that is consistent in detail with the paleoseismic evidence of the AD 1700 earthquake (McMillan and Hutchinson 2002).

This very brief review of the archeological record suggests there are several previously unrecognized links with the paleoenvironmental record. Discovering these correlations required development of new paleovegetation and paleoclimate records with accurate chronologies. For example, the major vegetation changes occurring at the time of the Manis site were not evident from the record at Manis (which was not ecotonal), but rather was evident only after publication of an accurate age of the Manis mastodon and development of detailed pollen records, such as at Yahoo Lake (Fig. 4.5). Thus, we expect additional refinement of chronologies and focus on high-resolution paleoenvironmental records to yield similar insights.

Chapter 6
Insights and Future Research Needs

Abstract The geography and environmental history of the Olympic Mountains arguably place the region as globally unique, and thus paleoecological insights from this region should also provide some new findings. The very steep precipitation gradient, the general lack of human influence on Holocene vegetation (as occurs in Europe), and the important role of an episodic fire regime allow for researchers to infer complex interactions within the scope of changing climate and disturbance regimes. The few ambiguities in species-level determination from the pollen flora and the low importance of *Pinus* (which swamps out other pollen taxa in drier climates of the Pacific Northwest), and the occurrence of leaf or needle macrofossils greatly aids the interpretation of the paleoecological record on the Olympic Peninsula. This chapter presents, in list form, the major conclusions emerging from the synthesis of paleoecological records. It then lists research questions that emerge from the current data, including both specific unresolved issues and how more dimensions of paleoenvironmental change may be addressed.

The modern environment of the Olympic Peninsula is marked by steep environmental gradients in precipitation and in temperature, resulting in a wide range of conditions (i.e., all four corners of the wet/dry and warm/cool environmental space) represented within a small region. Our review of vegetation gradients (Chap. 1) revealed the importance of spring snowpack in creating more distinct life-zone layering on the west side of the peninsula compared to the east side. The current trends in climate on the peninsula, especially the declining snowpack and the increase in temperature minima, therefore, should be affecting the ecotones among the vegetation zones. The paleoecological record, in conjunction with the paleoclimate record, is an obvious means of exploring how ecotone migration has occurred during past climate changes. Our motivation for this is paralleled by paleoecological studies of pollen and macrofossils in many other mountainous areas. Some of our take-home messages are in broad agreement with these other regions (Spear 1989; Jackson and Whitehead 1991; Tinner et al. 2005). These conclusions and insights are highlighted in this chapter, which we follow with research needs. We use a list form to clearly distinguish our conclusions:

- In the Olympic Mountains, climate change did not result in vertical displacement of compositionally constant vegetation zones. Elevational movement of

© Springer International Publishing Switzerland 2015

D. G. Gavin, L. B. Brubaker, *Late Pleistocene and Holocene Environmental Change on the Olympic Peninsula, Washington*, Ecological Studies 222,

DOI 10.1007/978-3-319-11014-1_6

vegetation zones is an attractive and well-worn concept that has been interpreted from paleorecords (e.g., van der Hammen 1974) or projected into the future (e.g., Franklin et al. 1992). However, it has long been recognized from studies of postglacial vegetation change that species respond individualistically to climate change, favoring a Gleasonian community concept (Davis 1986). Our results of modern analog analysis (Fig. 4.9) clearly show that simple vertical shifts were the exception rather than the rule. Vegetation modelers appreciate this issue and use models based on species traits rather than predicting community shifts. Recent studies that project climate impacts on vegetation use Dynamic Global Vegetation Models (DGVMs) and discuss results in terms of shifting species dominance rather than simple ecotone shifts (Halofsky et al. 2011).

- Vegetation change was characterized by dramatic changes in population size and spatial extent. Some species that are dominant today were very rare during the early Holocene (e.g., western hemlock) while other species that are largely confined to low or high elevation today (e.g., Sitka spruce and mountain hemlock) had a broader elevational range during the late-glacial. Complex terrain can support a complex set of microsites that allow the survival of small populations during periods of harsh climate. These small populations then can expand when more favorable conditions establish. The high rate of endemism, and bias of endemics to the riparian forests or dry subalpine meadows, is strongly suggestive of the importance of refugia during the postglacial period, and likely during the full-glacial period as well.

- Sites differ in their sensitivity to climate change. The coastal low-elevation sites consistently supported Sitka spruce, probably because seasonal temperature and moisture regimes on the coast were sufficiently maritime during the entire late-glacial to late Holocene. Subalpine fir persisted through the entire record at Moose, probably because that area was cold-dry during the entire late-glacial to late Holocene. In contrast, the site at the lower elevation of spring snowpack (Yahoo Lake) had the most sensitive response to climate change.

- All sites showed abrupt responses to climate change at the beginning of the Holocene, suggesting that despite the long life spans of the dominant tree species vegetation generally responded to climate change with minimal lag. It remains possible, however, that certain aspects of no-analog communities were the result of lagged responses to climate change. For example, the association of Sitka spruce and subalpine fir would be considered impossible based on modern plant geography on the peninsula, though it occurred twice at low elevations in the late-glacial and early Holocene. Similarly, the mapping of pollen data provided some evidence that Douglas-fir arrival to the region lagged behind climate change.

While there is a rich and detailed environmental history on the Olympic Peninsula based on sediment-based studies, there remains potential for improving our understanding in several areas. At the millennial scale, our results confirm previous studies regarding the general sequence of vegetation change at low elevations (Heusser 1977; Mathewes 1985; Whitlock 1992). Examination of mapped vegetation patterns and sub-millennial variability in the records raises many questions for future research. We provide these recommendations as a list, and their order is not meant to convey relative importance:

- Several specific aspects of vegetation history were unexpected and further investigation may reveal important aspects of vegetation–fire–climate interactions.

  - The chronologies of many sites, based on radiocarbon dates, limit the inferences that can be drawn from the regional patterns of forest composition in single time slices. For example, determining the synchroneity of late-glacial vegetation changes is limited by the difficulty to age organic-poor sediments. However, dating by accelerator mass spectrometry (AMS) methods, which requires as little as 50 mg of plant material, has become standard only within the past 15 years. Better-dated records are required to verify existing records and to accurately map the spatial spread of populations. In some cases, archived material exists from sediment cores that may be used to provide new chronologies on existing pollen records.
  - The outer west coast revealed less sensitivity to changing regional climate, perhaps due to the moderating effect of onshore winds (as initially proposed by Henry Hansen). However, upslope of the coastal plain in the Pacific silver fir zone, the forest experiences a very dynamic history. This increased sensitivity may be partly explained as an artifact of different sampling resolutions among sites. However, replicating sites with similar basin characteristics as at Yahoo Lake may result in records that are more comparable across the vegetation zones on the peninsula.
  - The Manis Mastodon site is world famous as a rare site demonstrating human hunting of megafauna. The late-glacial fauna of mastodon, elk, and bison ended during the late-glacial and by the early Holocene fossil assemblages resembled the modern fauna (Mustoe et al. 2005). The relative contribution of hunting versus climate change towards megafauna extinction remains debated. Studies that reconstruct the ecological changes during megafauna decline, and reconstruct the decline itself, have the potential to help understand herbivore effects on vegetation (Wilson et al. 2009) and, possibly, distinguish among the mechanisms causing the extinctions. Such a study in Indiana used the dung fungus *Sporormiella* to reconstruct declining megafauna abundance (Gill et al. 2009). The Sequim area and the Kitsap Peninsula contain numerous deep kettle lakes that may provide excellent undisturbed sediment for studying the relative sequences of vegetation change, faunal change, and forest fire.
  - Local nonclimatic site conditions might be important and override climatic controls of vegetation change. The slow colonization of Mountain hemlock zone forest (Fig. 4.6) presents a conundrum that requires more study. Was this site limited by soil texture, which required long periods of time and perhaps the addition of Mazama tephra to fill voids in talus slopes? How common was this on the peninsula, or in similar forests in the Cascade Range? What is the

implication for the development of meadows and subalpine forests during the early Holocene?

- The early Holocene landscape supported frequent fire and levels of Douglas-fir and red alder that approach what occurs on industrially managed forestland today. What was the spatial mosaic of this vegetation type? Did alder colonize hillslopes similar to what occurs today in the Oregon Coast Range? Has red alder greatly improved the nitrogen status of early Holocene soils, boosting forest productivity for subsequent millennia?

- The detailed paleoecological records spanning a climatic gradient provide an opportunity to use paleoecological records to calibrate vegetation models. Dynamic vegetation models, such as the Lund–Potsdam–Jena General Eco-system Simulator (LPJ-GUESS), can use climate inputs to predict the likely set of taxa on the landscape (e.g., Miller et al. 2008). Such an approach should be feasible considering the strong abiotic controls of modern vegetation on the peninsula and the strong climatic control of vegetation in the past. Our initial application of LPJ-GUESS, requiring some modifications to the model, is providing promising results showing a match between the pollen records and the simulated vegetation. Such a model could then be used to predict vegetation change under a trajectory of changing climate into the future. For example, two ocean–atmosphere general circulation models (CGCM3 and HADGEM1) run under the greenhouse gas scenario of >600 ppm CO_2 by the year 2080 are projecting a 3–5 °C increase in December–February mean temperature, a ca. 4 °C increase in June–August temperature, and a 15–40 % decrease in summer precipitation in an area including Yahoo Lake (Wang et al. 2012). Such a summer climate appears to be close to the warm and dry periods of the early Holocene at 10 ka, as inferred from the chironomid temperature reconstruction (Fig. 3.4). A rapid reversion of the forest to the early Holocene state would be dramatic. However, such a compositional shift will require the occurrence of large fires to overcome strong neighborhood effects on species composition. The 10-ka analog for future climates is crude because of differences in the seasonal climatic patterns between the two periods. The paleoclimate record indicates increased seasonality during the early Holocene (as seen from cooler winters inferred from the speleothem record). Given that winter temperatures are rising faster than summer temperatures, we are headed into a no-analog climate and future vegetation change is not likely to be similar to the early Holocene. Dynamic vegetation models are the current best available tools to project such change.

• There is potential to infer more dimensions of the paleoenvironment using interdiscisplinary biogeographic methods of modeling distributions, paleoecology, and genetics.

  - Can phylogeographic studies determine which of the endemic and disjunct taxa are likely to have been long-term ($>20,000$ years) residents of the peninsula? A common approach is to use molecular genetic approaches to show whether a disjunct population diverged from other populations (or sister taxa) before the Last Glacial Maximum (LGM), thus suggesting persistence in a glacial refugium (Hu et al. 2009). Very few such studies to our knowledge exist on the Olympic Peninsula.

- Can we improve upon the reconstructions of vegetation and climate during the glacial period? Most work has focused on lakes that formed during the initial glacial retreat around 14 ka. Cal Heusser's long pollen records from bogs, however, extend further back and provide an important view of the LGM, but hiatuses in the sediment and poor chronological control make interpretation difficult in many cases. As discussed above, there is potential to improve these long records with better radiocarbon dating and possibly dating to before 40 ka using optical spin luminescence.
- What is the climatic niche of the "paleoendemic" species? Does our understanding of past climate agree with their long-term persistence on the landscape? Can this approach provide new information regarding plasticity of the niche?

- Paleoecological studies may be designed to address knowledge gaps in specific settings.

- The prairies in Sitka spruce forests near Ozette, Forks, and Quillayute are widely regarded as anthropogenic, with significant ecological and ethnographic support for their origin and maintenance (Reagan 1909; Anderson 2009). However, their long-term presence and their biotic significance are not known. Radiocarbon dating soil charcoal and examining any stratigraphic record near these sites could provide a history of these sites that has the potential to be older than coastal sites that are lost from sea-level rise. Much work has begun on this topic (D. Conca, personal communication).
- Similarly, intentional burning by people for berries in subalpine meadows has not been studied on the Olympic Peninsula (see Lepofsky and Lertzman 2008).
- The elk glades in riparian forests may be a recent phenomenon due to high elk populations following wolf extirpation. What is the longer history of forest density in terrace forests of the Hoh, Bogachiel, Queets, and Quinault Rivers? Paleoecological studies, carefully done, could address this question though use of "small hollow" sediments that record local forest density. One such study on the Queets River showed the potential for this approach, but the study was not located in areas of high elk density (Greenwald and Brubaker 2001).
- Fire occurrence on the Olympic Peninsula is highly episodic. Fire periods in the past were unlike any in recent history. The dendrochronological record, much of it already collected, has much potential to describe the extent of fire on particular years or decades (Henderson and Peter 1981; Wendel and Zabowski 2010). Such studies could provide precedents for high-fire-risk years during severe drought.

Appendix

Appendix A: Treatment of Pollen Data for Production of Regional Maps

We obtained pollen data from the Neotoma Paleoecology Database (*http://www. neotomadb.org*), from published diagrams, and from unpublished data of the authors. We included all sites with a radiocarbon-based chronology from west of the Cascades and from the southern Puget Lowlands to southern Vancouver Island. If data were not available in digital format, we digitized the values from published pollen diagrams using GraphClick software.

Creating pollen maps requires quality control on each site contributing data. Without such assessment, outlier values may greatly affect interpretation at each site. For each of the 20 sites, we developed new relationships between depth in the sediment and age of the sediment (age models) by fitting a spline curve through all age-control points (calibrated radiocarbon dates, tephras, and core top dates). These relationships have already been described for the five sites on the Olympic Peninsula. We avoided using samples that required extrapolation below the lowest radiocarbon date. Top ages were based on the publication year of the paper if the coring date was not available.

Pollen percentages were expressed as a percentage of all arboreal pollen taxa (*Abies, Acer, Alnus, Betula, Chrysolepis/Lithocarpus, Thuja plicata, Fraxinus, Picea, Pinus, Populus, Pseudotsuga menziesii, Quercus, Tsuga heterophylla, Tsuga mertensiana*). Some pollen records differentiate *Alnus rubra* and *Alnus viridis* while others do not. For consistency across sites, we map the pollen type *Alnus*, combining the percentages of the two species when necessary. While traditional methods express pollen as a percentage of all terrestrial pollen types, our intent was to limit interpretation to the dominant tree species. Using percentages of arboreal taxa also avoids the issue of between-site variability of non-arboreal pollen taxa that could affect pollen abundance.

© Springer International Publishing Switzerland 2015
D. G. Gavin, L. B. Brubaker, *Late Pleistocene and Holocene Environmental Change on the Olympic Peninsula, Washington,* Ecological Studies 222,
DOI 10.1007/978-3-319-11014-1

For each pollen taxon, pollen percentages were estimated by linear interpolation between the samples straddling each of 20 time slices. Percentages were square-root transformed and symbolized to emphasize differences in low abundances. Basemaps of ice extent and coastlines are from Dyke et al. (2003). Data were interpolated and mapped with the R computing environment, which allowed simple updating of data.

Table A.1 Late Quaternary pollen records used for mapping pollen abundance. Site numbers refer to Fig. 4.11. Notes indicate special consideration of certain chronologies. See Table 2 for site locations

Site	Name	Source[a]	Chronology[b]	Other notes	References
25	Cedar Swamp	Author	B:3 A:3 T:1 D:1 (0.3–12.1)	Lower part of core not counted. Removed one bulk date due to AMS date at same level.	(McLachlan and Brubaker 1995)
24	Crocker Lake	Author	B:4 A:1 T:1 D:4 (0.03–13.2)	Tied pine decline to 11.6 ka and basal age tied to Cedar Swamp AMS age	(McLachlan and Brubaker 1995)
36	Davis Lake	Neotoma	B:16 T:1 (−0.01–7.1)	Did not use data below Mazama because 1) suspect hard-water effect on bulk dates in clay-rich sections, and 2) pollen differed greatly from nearby Mineral Lake	(Barnosky 1981)
18	East Sooke Fen	Neotoma	B:4 T:1 (0.2–13.5)		(Brown and Hebda 2002)
28	Hall Lake	Digitized	B:6 T:1 D:1 (0.03–15.2)	Radiocarbon times scale on diagram translated to depths, then to calibrated ages using new age model.	(Sugita and Tsukada 1982)
19	Kirk Lake	Neotoma	B:4 T:1 (−0.03–14.1)		(Cwynar 1987)
4	Marion Lake	Neotoma	B:7 T:1 (0.04–9.3)	Used Mike Lake for the period before 9.3 ka due to its stronger chronology. *Quercus* not in Neotoma database and therefore estimated from published diagram.	(Mathewes 1973; Mathewes and Heusser 1981)
29	Martins Lake	Author	A:2 T:1	Pine decline tied to 11.2 ka	(Gavin et al. 2001)

Table A.1 (continued)

Site	Name	Source[a]	Chronology[b]	Other notes	References
5	Mike Lake	Digitized	A:5[c] (9.9–13.8)	Several taxa absent from published diagram and not plotted on maps	(Pellatt et al. 2002)
35	Mineral Lake	Digitized	B:7 T:4[d] D:1 (−0.03–20.8)	Radiocarbon times scale on diagram translated to depths, then to calibrated ages using new age model.	(Sugita and Tsukada 1982)
26	Moose Lake	Author	B:4 A:1 T:1 D:3 (−0.01–14.3)	Pine decline tied to 11.6 ka	(Brubaker and McLachlan 1996; Gavin et al. 2001)
33	Nisqually Lake	Digitized	B:1 T:1 (0.04–14.9)		(Hibbert 1979)
12	Panther Potholes	Author	A:6 T:3 (−0.04–10.6)		(Prichard et al. 2009)
2	Pinecrest Lake	Neotoma	B:1 T:1 (0.04–12.9)		(Mathewes et al. 1972; Mathewes and Rouse 1975)
16	Pixie Lake	Neotoma	B:7 T:1 (0.02–15.6)	Did not extrapolate below lowest ^{14}C age	(Brown and Hebda 2002; Brown and Hebda 2003)
8	Porphyry Lake	Neotoma	B:4 (0.30–14.7)	Did not extrapolate below lowest ^{14}C age	(Brown and Hebda 2003)
17	Walker Lake	Neotoma	B:4 T:1 (0.03–14.2)	Did not extrapolate below lowest ^{14}C age	(Brown and Hebda 2003)
23	Wentworth Lake	Author	A:5 T:1 (0.3–13.9)		This study
11	Whyac Lake	Neotoma	B:4 (0.3–12.7)	Did not extrapolate below lowest ^{14}C age	(Brown and Hebda 2002; Brown and Hebda 2003)
30	Yahoo Lake	Author	A:5 T:1 (0.6–14.6)		Gavin et al. 2013

[a] "Neotoma" is the Neotoma Paleoecology Database. "Digitized" indicates values were hand-digitized from pollen diagram graphics using GraphClick software. "Author" indicates data were provided directly by researchers

[b] B number of bulk (conventional) radiocarbon dates, A number of AMS (accelerator mass spectrometry) radiocarbon dates, T number of tephras with a known age, D number dates not used because deemed unreliable. Values in parentheses indicate total age range used in maps (in ka, thousands of calibrated years before present)

[c] Four dates on pollen concentrates, one on a needle

[d] Three tephra ages not used (insufficient information to date)

Literature

Acker SA, Beechie TJ, Shafroth PB (2008) Effects of a natural dam-break flood on geomorphology and vegetation on the Elwha River, Washington, USA. Northwest Sci 82:210–223. doi: 10.3955/0029-344X-82.S.I.210

Adams MJ, Bury RB (2002) The endemic headwater stream amphibians of the American Northwest: associations with environmental gradients in a large forested preserve. Global Ecol Biogeogr 11:169–178. doi:10.1046/j.1466-822X.2002.00272.x

Adams RP, Hunter G, Fairhall TA (2010) Discovery and SNPs analyses of populations of *Juniperus maritima* in the Olympic Peninsula, a Pleistocene refugium? Phytologia 92:68–81

Agee JK (1993) Fire ecology of Pacific Northwest forests, 1st edn. Island Press, Washington, DC

Agee JK, Flewelling R (1983) A fire cycle model based on climate for the Olympic Mountains, Washington. Seventh conference: Fire and forest meteorology conference, Fort Collins, CO. American Meteorological Society, Boston, MA, pp 32–37

Agee JK, Huff MH (1987) Fuel succession in a western hemlock/Douglas-fir forest. Can J Forest Res 17:697–704. doi:10.1139/x87-112

Agee JK, Smith L (1984) Subalpine tree reestablishment after fire in the Olympic Mountains, Washington. Ecology 65:810–819. doi:10.2307/1938054

Allen GA (2001) Hybrid speciation in Erythronium (Liliaceae): a new allotetraploid species from Washington State. Syst Bot 26:263–272. doi: 10.1043/0363-6445-26.2.263

Ames KM (2005) The place of Ozette in Northwest Coast Archaeology. Ozette Archaeological project research reports Volume III: Ethnobotany and wood technology. WSU Department of Anthropology Reports of Research, Seattle, pp 9–24

Anders AM, Roe GH, Durran DR, Minder JR (2007) Small-scale spatial gradients in climatological precipitation on the Olympic Peninsula. J Hydrometeorol 8:1068–1081. doi:10.1175/JHM610.1

Anderson KH (2009) The Ozette prairies of Olympic National Park: their former indigenous uses and management. Final report to Olympic National Park, Port Angeles

Ashworth AC (2003) Quaternary Coleoptera of the United States and Canada. In: Gillespie AR, Porter SC, Atwater BF (eds) The Quaternary period in the United States, 1st edn. Elsevier, Amsterdam, pp 505–518

Ashworth AC, Nelson RE (2014) The paleoenvironment of the Olympia beds based on fossil beetles from Discovery Park, Seattle, Washington, U.S.A. Quat Int 341:243–254. doi:10.1016/j.quaint.2013.09.022

Ashworth AC, Gutenkunst M, Nelson RE (2000) The climate during the last advance of the Puget Sound glacier (Vashon Stade, Fraser glaciation) based on fossil beetles from Seattle. Abstracts with Program. Geological Society of America, Reno, p A-404

Atwater BF (1987) Evidence for great Holocene earthquakes along the outer coast of Washington State. Science 236:942–944. doi:10.1126/science.236.4804.942

© Springer International Publishing Switzerland 2015
D. G. Gavin, L. B. Brubaker, *Late Pleistocene and Holocene Environmental Change on the Olympic Peninsula, Washington,* Ecological Studies 222,
DOI 10.1007/978-3-319-11014-1

Atwater BF, Yamaguchi DK (1991) Sudden, probably coseismic submergence of Holocene trees and grass in coastal Washington State. Geology 19:706–709. doi:10.1130/0091-7613(1991)019<0706:SPCSOH>2.3.CO;2

Atwater BF, Musumi-Rokkaku S, Satake K et al (2005) The Orphan Tsunami of 1700: Japanese clues to a parent earthquake in North America. University of Washington Press/U.S. Geological Survey Professional Paper 1707.

Babcock RS, Burmester RF, Engebretson DC et al (1992) A rifted margin origin for the crescent basalts and related rocks in the northern Coast Range Volcanic Province, Washington and British Columbia. J Geophys Res: Solid Earth 97:6799–6821. doi:10.1029/91JB02926

Barnosky C (1981) A record of late Quaternary vegetation from Davis Lake, southern Puget Lowland, Washington. Quat Res 16:221–239. doi:10.1016/0033-5894(81)90046-6

Barnosky CW (1984) Late Pleistocene and early Holocene environmental history of southwestern Washington State, U.S.A. Can J Earth Sci 21:619–629. doi:10.1139/e84-068

Barr CB (2011) *Bryelmis* Barr (Coleoptera: Elmidae: Elminae), a new genus of riffle beetle with three new species from the Pacific Northwest, U.S.A. Coleopt Bull 65:197–212. doi:10.1649/072.065.0301

Barron JA, Heusser L, Herbert T, Lyle M (2003) High-resolution climatic evolution of coastal northern California during the past 16,000 years. Paleoceanography 18:1020. doi:10.1029/2002PA000768

Bartlein PJ, Harrison SP, Brewer S et al (2010) Pollen-based continental climate reconstructions at 6 and 21 ka: a global synthesis. Clim Dyn 37:775–802. doi:10.1007/s00382-010-0904-1

Belsky J, del Moral R (1982) Ecology of an alpine–subalpine meadow complex in the Olympic Mountains, Washington. Can J Bot 60:779–788. doi: 10.1139/b82-101

Bergland EO (1983) Prehistory and ethnography: Olympic National Park. National Park Service Pacific Northwest Region Division of Cultural Resources. Seattle

Beschta RL, Ripple WJ (2008) Wolves, trophic cascades, and rivers in the Olympic National Park, USA. Ecohydrology 1:118–130. doi:10.1002/eco.12

Blois JL, Williams JW, Fitzpatrick MC et al (2013) Modeling the climatic drivers of spatial patterns in vegetation composition since the last Glacial Maximum. Ecography 36:460–473. doi:10.1111/j.1600-0587.2012.07852.x

Booth DB, Troost KG, Clague JJ, Waitt RB (2003) The Cordilleran ice sheet. In: Gillespie AR, Porter SC, Atwater BF (eds) The Quaternary period in the United States, 1st edn. Elsevier, Amsterdam, pp 17–43

Bortenschlager S (1990) Aspects of pollen morphology in the Cupressaceae. Grana 29:129–138. doi:10.1080/00173139009427743

Brewer MC, Mass CF, Potter BE (2012) The west coast thermal trough: climatology and synoptic evolution. Mon Weather Rev 140:3820–3843. doi:10.1175/MWR-D-12-00078.1

Brown TA (1994) Radiocarbon dating of pollen by accelerator mass spectrometry. Ph. D. Thesis, University of Washington

Brown KJ, Hebda RJ (2002) Origin, development, and dynamics of coastal temperate conifer rainforests of southern Vancouver Island, Canada. Can J Forest Res 32:353–372. doi: 10.1139/x01-197

Brown KJ, Hebda RJ (2003) Coastal rainforest connections disclosed through a Late Quaternary vegetation, climate, and fire history investigation from the Mountain Hemlock Zone on southern Vancouver Island, British Colombia, Canada. Rev Palaeobot Palynol 123:247–269. doi: 10.1016/S0034-6667(02)00195-1

Brubaker LB (1991) Climate change and the origin of old-growth Douglas-fir forests in the Puget Sound Lowland. In: Ruggiero LF, Aubry KB, Carey AB, Huff MH (eds) Wildlife and vegetation of unmanaged Douglas-fir forests. USDA Forest Service, Pacific Northwest Research Station, Portland, pp 17–24

Brubaker LB, McLachlan JS (1996) Landscape diversity and vegetation response to long-term climate change in the eastern Olympic Peninsula, Pacific Northwest, USA. In: Walker B, Steffen W (eds) Global change and terrestrial ecosystems. Cambridge University Press, New York, pp 184–203

Buckingham NM, Schreiner EG, Kaye TN et al (1995) Flora of the Olympic Peninsula, Washington. Northwest Interpretative Association, Seattle

Burke TE (2005) Conservation assessment for four species of the genus Hemphillia. Unpublished report, USDA Forest Service Region 6, Portland

Butler VL, Campbell SK (2004) Resource intensification and resource depression in the Pacific Northwest of North America: a zooarchaeological review. J World Prehist 18:327–405. doi:10.1007/s10963-004-5622-3

Causey NB (1954) New records and species of millipedes from the western United States and Canada. Pan-Pac Entomol 30:221–227

Chase M, Bleskie C, Walker I et al (2008) Midge-inferred Holocene summer temperatures in Southeastern British Columbia, Canada. Palaeogeogr Palaeoclimatol Palaeoecol 257:244–259. doi:10.1016/j.palaeo.2007.10.020

Clague JJ, Mathewes RW, Guilbault JP et al (1997) Pre-Younger Dryas resurgence of the southwestern margin of the Cordilleran ice sheet, British Columbia, Canada. Boreas 26:261–277. doi: 10.1111/j.1502-3885.1997.tb00855.x

COHMAP Members (1988) Climatic changes of the last 18,000 years: observations and model simulations. Science 241:1043–1052. doi:10.1126/science.241.4869.1043

Compton JE, Church MR, Larned ST, Hogsett WE (2003) Nitrogen export from forested watersheds in the Oregon Coast Range: The role of N2-fixing red alder. Ecosystems 6:773–785. doi:10.1007/s10021-002-0207-4

Cong S (1997) Fossils of an undescribed blind trechine (Coleoptera: Carabidae) from near Kalaloch, Olympic Peninsula, Washington. Coleopt Bull 51:208–211. doi:10.2307/4009411

Cong SG, Ashworth AC (1996) Palaeoenvironmental interpretation of Middle and Late Wisconsinan fossil Coleopteran assemblages from western Olympic Peninsula, Washington, USA. J Quat Sci 11:345–356. doi:10.1002/(SICI)1099-1417(199609/10)11:5<345::AID-JQS259>3.0.CO;2-A

Conway H, Rasmussen LA, Marshall H-P (1999) Annual mass balance of Blue Glacier, USA: 1955–97. Geogr Ann Ser A, Phys Geogr 81:509–520

Cook RE (1969) Variation in species density of North American birds. Syst Zool 18:63–84. doi:10.2307/2412411

Croes DR (1989) Prehistoric ethnicity on the northwest coast of North America: an evaluation of style in basketry and lithics. J Anthropol Archaeol 8:101–130

Croes DR, Hackenberger S (1988) Hoko River archeological complex: modeling prehistoric Northwest Coast economic evolution. Res Econ Anthropol suppl 3:19–86

Cwynar LC (1987) Fire and forest history of the North Cascade Range. Ecology 68:791–802. doi:10.2307/1938350

Cwynar L, Burden E, McAndrews J (1979) Inexpensive sieving method for concentrating pollen and spores from fine-grained sediments. Can J Earth Sci 16:1115–1120. doi:10.1139/e79-097

Dale VH, Hemstrom M, Franklin J (1986) Modeling the long-term effects of disturbances on forest succession, Olympic Peninsula, Washington. Can J Forest Res 16:56–67. doi:10.1139/x86-010

Daly C, Gibson WP, Taylor GH et al (2002) A knowledge-based approach to the statistical mapping of climate. Clim Res 22:99–113. doi:10.3354/cr022099

Davis MB (1986) Climatic instability, time lags, and community disequilibrium. In: Diamond J, Chase TJ (eds) Community Ecology. Harper and Row, New York, pp 269–284

Deevey ES (1948) Review of postglacial forest succession, climate, and chronology in the Pacific Northwest by Henry P. Hansen. Geogr Rev 38:345–347. doi:10.2307/210873

DeHaan P, Adams B, Hawkins D et al (2013) Analysis of genetic variaion in the Olympic Mudminnow (Novumbra hubbsi). Final report, US Fish and Wildlife Service

Del Moral R (1984) The impact of the Olympic marmot on subalpine vegetation structure. Am J Bot 71:1228–1236

Delcourt HR, Delcourt P (1991) Quaternary ecology: a paleoecological perspective. Springer, New York

Demboski JR, Sullivan J (2003) Extensive mtDNA variation within the yellow-pine chipmunk, Tamias amoenus (Rodentia: Sciuridae), and phylogeographic inferences for northwest North America. Mol Phylogenet Evol 26:389–408. doi:10.1016/S1055-7903(02)00363-9

Deser C, Phillips AS, Hurrell JW (2004) Pacific interdecadal climate variability: linkages between the Tropics and the North Pacific during Boreal Winter since 1900. J Clim 17:3109–3124. doi:10.1175/1520-0442 (2004) 017<3109:PICVLB>2.0.CO;2

Devine W, Aubry C, Bower A et al (2012) Climate change and forest trees in the Pacific Northwest: a vulnerability assessment and recommended actions for national forests. U.S. Department of Agriculture, Forest Service, Pacific Northwest Region, Olympia

Dolédec S, Chessel D, Braak CJF ter, Champely S (1996) Matching species traits to environmental variables: a new three-table ordination method. Environ Ecol Stat 3:143–166. doi:10.1007/BF02427859

Dunwiddie PW (1985) Dichotomous key to conifer foliage in the Pacific Northwest. Northwest Sci 59:185–191

Dunwiddie PW (1986) A 6000-year record of forest history on Mount Rainier, Washington. Ecology 67:58–68. doi:10.2307/1938503

Dunwiddie PW (1987) Macrofossil and pollen representation of coniferous trees in modern sediments from Washington. Ecology 68:1–11. doi:10.2307/1938800

Dyke AS, Moore A, Robertson L (2003) Deglaciation of North America. Open file 1574 Geological Survey of Canada

Easterbrook DJ (2004) Comments on "Implications of a late-glacial pollen record for the glacial and climatic history of the Fraser Lowland, British Columbia" by Pellatt et al., 2002. Palaeogeogr Palaeoclimat Palaeoecol 203:337–342. doi:10.1016/S0031-0182(03)00722-3

Edelman AJ (2003) Marmota olympus. Mamm Species 736:1–5. doi:10.1644/736

Eidenshink JC, Schwind B, Brewer K et al (2007) A project for monitoring trends in burn severity. Fire Ecol 3:3–21. doi:IND44041099

Elias SA (2013) The problem of conifer species migration lag in the Pacific Northwest region since the last glaciation. Quat Sci Rev 77:55–69. doi:10.1016/j.quascirev.2013.07.023

Ersek V, Clark PU, Mix AC et al (2012) Holocene winter climate variability in mid-latitude western North America. Nat Commun. doi:10.1038/ncomms2222

Ettinger AK, Ford KR, HilleRisLambers J (2011) Climate determines upper, but not lower, altitudinal range limits of Pacific Northwest conifers. Ecology 92:1323–1331. doi:10.1890/10-1639.1

Ettl GJ, Peterson DL (1995) Growth response of subalpine fir (*Abies lasiocarpa*) to climate in the Olympic Mountains, Washington, USA. Global Change Biol 1:213–230. doi:10.1111/j.1365-2486.1995.tb00023.x

Faegri K, Iversen J (2000) Textbook of pollen analysis, 4th edn. Blackburn Press, Caldwell

Fahnestock GR, Agee JK (1983) Biomass consumption and smole production by prehistoric and modern forest fires in western Washington. J Forest 81:653–657

Fedje DW, Sumpter ID, Southon JR (2009) Sea-levels and archaeology in the Gulf Islands National Park Reserve. Can J Archaeol 33:234–253

Fedje D, Mackie Q, Lacourse T, McLaren D (2011) Younger Dryas environments and archaeology on the Northwest Coast of North America. Quat Int 242:452–462. doi:10.1016/j.quaint.2011.03.042

Fisher DM (2013) Postglacial transient dynamics of Olympic Peninsula Forests: comparing predictions and observations. M.S. Thesis, University of Oregon

Fonda RW (1974) Forest succession in relation to river terrace development in Olympic National Park, Washington. Ecology 55:927–942. doi:10.2307/1940346

Fonda RW, Binney EP (2011) Vegetation response to prescribed fire in Douglas-fir forests, Olympic National Park. Northwest Sci 85:30–40. doi:10.3955/046.085.0103

Fonda RW, Bliss LC (1969) Forest vegetation of the montane and subalpine zones, Olympic Mountains, Washington. Ecol Monogr 39:271–301. doi:10.2307/1948547

Forest Health Program (2008) Tree blowdown aerial survey: March 6–7 2008, Washington Coast. Washington State Department of Natural Resources. http://www.dnr.wa.gov/Publications/rp_fh_2008_blowdownreport.pdf. Accessed 6 Nov 2012

Franklin JE, Dyrness CT (1988) Natural vegetation of Oregon and Washington. Oregon State University Press, Oregon

Franklin JF, Swanson FJ, Harmon ME et al (1992) Effects of global climate change on forests in northwestern North America. In: Peters RL, Lovejoy TE (eds) Global warming and biologic diversity. Yale University Press, New Haven, pp 244–258

Friele PA, Clague JJ (2002) Younger Dryas readvance in Squamish River valley, southern Coast Mountains, British Columbia. Quat Sci Rev 21:1925–1933. doi:10.1016/S0277-3791(02)00081-1

Gavin DG (2009) The coastal-disjunct mesic flora in the inland Pacific Northwest of USA and Canada: refugia, dispersal and disequilibrium. Diversity Distribut 15:972–982. doi:10.1111/j.1472-4642.2009.00597.x

Gavin DG, Brubaker LB (1999) A 6000-year soil pollen record of subalpine meadow vegetation in the Olympic Mountains, Washington, USA. J Ecol 87:106–122. doi:10.1046/j.1365-2745.1999.00335.x

Gavin DG, Hu FS (2006) Spatial variation of climatic and non-climatic controls on species distribution: the range limit of *Tsuga heterophylla*. J Biogeogr 33:1384–1396. doi:10.1111/j.1365-2699.2006.01509.x

Gavin DG, McLachlan JS, Brubaker LB, Young KA (2001) Postglacial history of subalpine forests, Olympic Peninsula, Washington, USA. Holocene 11:177–188. doi:10.1191/095968301670879949

Gavin DG, Brubaker LB, Lertzman KP (2003a) Holocene fire history of a coastal temperate rain forest based on soil charcoal radiocarbon dates. Ecology 84:186–201. doi:10.1890/0012-9658(2003)084[0186:HFHOAC]2.0.CO;2

Gavin DG, Brubaker LB, Lertzman KP (2003b) An 1800-year record of the spatial and temporal distribution of fire from the west coast of Vancouver Island, Canada. Can J Forest Res 33:573–586. doi:10.1139/X02-196

Gavin DG, Brubaker LB, McLachlan JS, Oswald WW (2005) Correspondence of pollen assemblages with forest zones across steep environmental gradients, Olympic Peninsula, Washington, USA. Holocene 15:648–662. doi:10.1191/0959683605hl841rp

Gavin DG, Hallett DJ, Hu FS et al (2007) Forest fire and climate change in western North America: insights from sediment charcoal records. Front Ecol Environ 5:499–506. doi:10.1890/060161

Gavin DG, Henderson ACG, Westover K et al (2011) Abrubt Holocene climate change and potential response to solar forcing in western Canada. Quat Sci Rev 30:1243–1255. doi:10.1016/j.quascirev.2011.03.003

Gavin DG, Brubaker LB, Greenwald DN (2013) Postglacial climate and fire-mediated vegetation change on the western Olympic Peninsula, Washington (USA). Ecol Monogr 83:471–489. doi:10.1890/12-1742.1

Gedalof Z, Smith DJ (2001) Dendroclimatic response of mountain hemlock (*Tsuga mertensiana*) in Pacific North America. Can J Forest Res 31:322–332. doi:10.1139/x00-169

Geertsema M, Pojar JJ (2007) Influence of landslides on biophysical diversity—A perspective from British Columbia. Geomorphology 89:55–69. doi:10.1016/j.geomorph.2006.07.019

Gerstel WJ (1999) Landslide inventory of the west-central Olympic Peninsula. Open File Report 99-2 Washington Division of Geology and Earth Resources, Olympia

Gilbert MTP, Jenkins DL, Götherstrom A et al (2008) DNA from pre-Clovis human coprolites in Oregon, North America. Science 320:786–789. doi:10.1126/science.1154116

Gill JL, Williams JW, Jackson ST et al (2009) Pleistocene megafaunal collapse, novel plant communities, and enhanced fire regimes in North America. Science 326:1100–1103. doi:10.1126/science.1179504

Goldner A, Herold N, Huber M (2013) The challenge of simulating warmth of the mid-Miocene climate optimum in CESM1. Clim Past Discuss 9:3489–3518. doi:10.5194/cpd-9-3489-2013

Graham A (2011) A natural history of the new world: the ecology and evolution of plants in the Americas. University of Chicago Press, Chicago

Greenwald DN, Brubaker LB (2001) A 5000-year record of disturbance and vegetation change in riparian forests of the Queets River, Washington, USA. Can J Forest Res 31:1375–1385

Grier C (2003) Dimensions of regional interaction in the prehistoric Gulf of Georgia. In: Matson RG, Coupland G, Mackie Q (eds) Emerging from the mist. Studies in Northwest Coast culture history. University of British Columbia Press, Vancouver, pp 170–187

Gugger PF, Sugita S (2010) Glacial populations and postglacial migration of Douglas-fir based on fossil pollen and macrofossil evidence. Quat Sci Rev 29:2052–2070. doi:10.1016/j.quascirev.2010.04.022

Gustafson CE, Gilbow DW, Daugherty RD (1979) The Manis mastodon site: early man on the Olympic Peninsula. Can J Archaeol 3:157–164. doi:10.2307/41102203

Gutchwesky M (2004) A paleoenvironmental reconstruction of the Ozette Prairies from analysis of peatland cores, Olympic National Park, Washington. M.A. Thesis, Western Washington University

Hall ER (1945) Four new ermines from the Pacific Northwest. J Mammal 26:75–85. doi:10.2307/1375034

Hallett DJ, Lepofsky DS, Mathewes RW, Lertzman KP (2003) 11,000 years of fire history and climate in the mountain hemlock rain forests of southwestern British Columbia based on sedimentary charcoal. Can J Forest Res 33:292–312. doi:10.1139/x02-177

Halofsky JE, Peterson DL, O'Halloran KA, Hawkins Hoffman C (eds) (2011) Adapting to climate change at Olympic National Forest and Olympic National Park. U S Forest Service General Technical Report PNW 844

Hampe A, Jump AS (2011) Climate relicts: past, present, future. Annu Rev Ecol Evol Syst 42:313–333. doi:10.1146/annurev-ecolsys-102710-145015

Hansen HP (1941) Paleoecology of a bog in the spruce-hemlock climax of the Olympic Peninsula. Am Midl Nat 25:290–297

Hansen HP (1947) Postglacial forest succession, climate, and chronology in the Pacific Northwest. American Philosophical Society, Philadelphia

Hansen BCS, Easterbrook DJ (1974) Stratigraphy and palynology of late Quaternary sediments in the Puget Lowland, Washington. Geol Soc Am Bull 85:587–602. doi:10.1130/0016-7606(1974)85<587:SAPOLQ>2.0.CO;2

Hansen B, Engstrom D (1996) Vegetation history of Pleasant Island, southeastern Alaska, since 13,000 year BP. Quat Res 46:161–175. doi:10.1006/qres.1996.0056

Harington CR (1975) Pleistocene Muskoxen (Symbos) from Alberta and British Columbia. Can J Earth Sci 12:903–919. doi:10.1139/e75-083

Harington CR (2008) The evolution of arctic marine mammals. Ecol Appl 18:S23–S40. doi:10.1890/06-0624.1

Harmon ME, Franklin JF (1983) Age distribution of western hemlock and its relation to Roosevelt Elk populations in the South Fork Hoh River Valley, Washington. Northwest Sci 57:249–255

Harmon ME, Franklin JF (1989) Tree seedlings on logs in Picea-Tsuga forests of Oregon and Washington. Ecology 70:48–59. doi:10.2307/1938411

Harrington CA, Gould PJ (2010) Growth of western redcedar and yellow-cedar. In: Harrington CA (ed) A tale of two cedars: international symposium on Western redcedar and yellow-cedar. GTR-PNW-828. U.S. Department of Agriculture, Forest Service, Pacific Northwest Research Station, Portland, pp 97–102

Harvey AE, Byler JW, McDonald GI et al (2008) Death of an ecosystem: perspectives on western white pine ecosystems of North America at the end of the twentieth century. USDA Forest Service, Rocky Mountain Research Station, Fort Collins

Hebda RJ, Mathewes RW (1984) Holocene history of cedar and native Indian cultures of the North American Pacific Coast. Science 225:711–713

Heine JT (1998) Extent, timing, and climatic implications of glacier advances Mount Rainier, Washington, U.S.A., at the Pleistocene/Holocene transition. Quat Sci Rev 17:1139–1148. doi:10.1016/S0277-3791(97)00077-2

Heinselman M (1973) Fire in the virgin forests of the Boundary Waters Canoe Area, Minnesota. Quat Res 3:329–382

Hellwig J (2010) The interaction of climate, tectonics, and topography in the Olympic Mountains of Washington State: the influence of erosion on tectonic steady-state and the synthesis of the Alpine Glacial history. M.S. Thesis, University of Illinois

Hemstrom MA, Logan SE (1986) Plant association and management guide Siuslaw National Forest, p 125

Henderson J, Peter D (1981) Preliminary plant associations and habitat types of the Shelton Ranger District, Olympic National Forest, p 53

Henderson JA, Peter DH, Lesher RD, Shaw DC (1989) Forested plant associations of the Olympic National Forest. USDA Forest Service R6 ECOL Technical Paper 001-88, Portland

Henderson JA, Lesher RD, Peter DH, Ringo CD (2011) A landscape model for predicting potential natural vegetation of the Olympic Peninsula USA using boundary equations and newly developed environmental variables. USDA Forest Service General Technical Report PNW-GTR-941, Portland

Heusser CJ (1957) Variations of Blue, Hoh, and White Glaciers during recent centuries. Arctic 10:139–150

Heusser CJ (1960) Late-pleistocene environments of North Pacific North America. American Geographical Society, New York

Heusser CJ (1964) Palynology of four bog sections from the western Olympic Peninsula, Washington. Ecology 45:23–40. doi:10.2307/1937104

Heusser CJ (1969) Modern pollen spectra from the Olympic Peninsula, Washington. Bull Torrey Bot Club 96:407–417

Heusser CJ (1972) Palynology and phytogeographical significance of a late Pleistocene refugium near Kalaloch Washington. Quat Res 2:189–201. doi:10.1016/0033-5894(72)90038-5

Heusser CJ (1973) Environmental sequence following the Fraser advance of the Juan de Fuca Lobe, Washington. Quat Res 3:284–306. doi:10.1016/0033-5894(73)90047-1

Heusser CJ (1974) Quaternary vegetation, climate, and glaciation of Hoh River Valley, Washington. Geol Soc Am Bull 85:1547–1560. doi:10.1130/0016-7606 (1974) 85<1547:QVCAGO>2.0.CO;2

Heusser CJ (1977) Quaternary palynology of the Pacific slope of Washington. Quaternary Research 8:282–306. doi:10.1016/0033-5894(77)90073-4

Heusser CJ (1978) Palynology of Quaternary deposits of the lower Bogachiel River area, Olympic Peninsula, Washington. Can J Earth Sci 15:1568–1578. doi:10.1139/e78-162

Heusser CJ, Heusser LE, Streeter SS (1980) Quaternary temperatures and precipitation for the northwest coast of North America. Nature 286:702–704. doi:10.1038/286702a0

Heusser CJ, Heusser LE, Peteet DM (1999) Humptulips revisited: a revised interpretation of Quaternary vegetation and climate of western Washington, USA. Palaeogeogr Palaeoclimatol Palaeoecol 150:191–221. doi:10.1016/S0031-0182(98)00225-9

Hibbert DM (1979) Pollen analysis of late quaternary sediments from two lakes in the Southern Puget Lowland, Washington. M.S. Thesis, University of Washington

Hicock SR, Hebda RJ, Armstrong JE (1982) Lag of the Fraser glacial maximum in the Pacific Northwest: pollen and macrofossil evidence from western Fraser Lowland, British Columbia. Can J Earth Sci 19:2288–2296. doi:10.1139/e82-201

Higuera PE, Gavin DG, Bartlein PJ, Hallett DJ (2010) Peak detection in sediment-charcoal records: impacts of alternative data analysis methods on fire-history interpretations. Int J Wildland Fire 19:996–1014. doi:10.1071/WF09134

Holliday JA, Yuen M, Ritland K, Aitken SN (2010) Postglacial history of a widespread conifer produces inverse clines in selective neutrality tests. Mol Ecol 19:3857–3864. doi:10.1111/j.1365-294X.2010.04767.x

Holman ML, Peterson DL (2006) Spatial and temporal variability in forest growth in the Olympic Mountains, Washington: sensitivity to climatic variability. Can J Forest Res 36:92–104. doi:10.1139/X05-225

Houston DB, Schreiner EG, Buckingham NM (1994) Biogeography of the Olympic Peninsula. In: Houston DB, Schreiner EG, Moorhead BB (eds) Mountain goats in Olympic National Park: biology and management of an introduced species

Hu FS, Hampe A, Petit RJ (2009) Paleoecology meets genetics: deciphering past vegetational dynamics. Front Ecol Environ 7:371–379. doi:10.1890/070160

Huff MH (1995) Forest age structure and development following wildfires in the western Olympic Mountains, Washington. Ecol Appl 5:471–483. doi:10.2307/1942037

Hutchinson I, McMillan AD (1997) Archaeological evidence for village abandonment asssociated with Late Holocene earthquakes at the northern Cascadia subduction zone. Quat Res 48:79–87. doi:10.1006/qres.1997.1890

Hutten KM, Torgersen CE, Woodward A et al (2012) Landscape patterns of balsam wooly adelgid occurrence and subalpine fir mortality on the Olympic Peninsula. Proceedings of the 59th annual western international forest disease work conference. WIFDWC, Leavenworth, pp 117–121

Impara PC (1997) Spatial and temporal patterns of fire in the forests of the Central Oregon Coast Range. Ph.D., Oregon State University

Jackson ST, Whitehead DR (1991) Holocene vegetation patterns in the Adirondack Mountains. Ecology 72:641–653. doi:10.2307/2937204

Jakob M (2000) The impacts of logging on landslide activity at Clayoquot Sound, British Columbia. Catena 38:279–300. doi:10.1016/S0341-8162(99)00078-8

James T, Gowan EJ, Hutchinson I et al (2009) Sea-level change and paleogeographic reconstructions, southern Vancouver Island, British Columbia, Canada. Quat Sci Rev 28:1200–1216. doi:10.1016/j.quascirev.2008.12.022

Jenkins K, Woodward A, Schreiner EG (2002) A framework for long-term ecological monitoring in Olympic National Park: prototype for the coniferous forest biome. Forest and Rangeland Ecosystem Science Center Olympic Field Station

Jenkins DL, Davis LG, Stafford TW et al (2012a) Clovis age western stemmed projectile points and human coprolites at the Paisley Caves. Science 337:223–228. doi:10.1126/science.1218443

Jenkins KJ, Happe PJ, Beirne KF, et al. (2012b) Recent population trends of mountain goats in the Olympic Mountains, Washington. Northwest Sci 86:264–275. doi:10.3955/046.086.0403

Kavanaugh DH (1979) Studies on the Nebriini (Coleoptera: Carabidae), III: New Nearctic Nebria species and subspecies, nomenclatural notes, and lectotype designations. Proc California Acad Sci 42:87–133

Kavanaugh DH (1981) Studies on the Nebriini (Coleoptera: Carabidae). IV. Four new Nebria taxa from western North America. Proc California Acad Sci 42:435–442

Kenady SM, Wilson MC, Schalk RF, Mierendorf RR (2011) Late Pleistocene butchered Bison antiquus from Ayer Pond, Orcas Island, Pacific Northwest: age confirmation and taphonomy. Quat Int 233:130–141. doi:10.1016/j.quaint.2010.04.013

Keppel G, Van Niel KP, Wardell-Johnson GW et al (2012) Refugia: identifying and understanding safe havens for biodiversity under climate change. Global Ecol Biogeogr 21:393–404. doi:10.1111/j.1466-8238.2011.00686.x

Kienast SS, McKay JL (2001) Sea surface temperatures in the subarctic Northeast Pacific reflect millennial-scale climate oscillations during the last 16 kyrs. Geophys Res Lett 28:1563–1566. doi:10.1029/2000GL012543

Kirk R, Daugherty RD (2007) Archaeology in Washington. University of Washington Press, Seattle

Kiss GK (1989) Engelmann × Sitka spruce hybrids in central British Columbia. Can J For Res 19:1190–1193. doi:10.1139/x89-178

Kitzberger T, Aráoz E, Gowda JH et al (2012) Decreases in fire spread probability with forest age promotes alternative community states, reduced resilience to climate variability. Ecosystems 15:97–112. doi:10.1007/s10021-011-9494-y

Knapp PA, Hadley KS (2012) A 300-year history of Pacific Northwest windstorms inferred from tree rings. Glob Planet Change 92-93:257–266. doi:10.1016/j.gloplacha.2012.06.002

Kovanen DJ, Easterbrook DJ (2002) Timing and Extent of Allerød and Younger Dryas Age (ca. 12,500-10,000 14C yr B.P.) Oscillations of the Cordilleran ice sheet in the Fraser Lowland, Western North America. Quat Res 57:208–224. doi:10.1006/qres.2001.2307

Kuramoto RT, Bliss LC (1970) Ecology of subalpine meadows in the Olympic Mountains, Washington. Ecol Monogr 40:317–347. doi:10.2307/1942286

Kutzbach J, Gallimore R, Harrison S et al. (1998) Climate and biome simulations for the past 21,000 years. Quat Sci Revi 17:473–506. doi:10.1016/S0277-3791(98)00009-2

Lacourse T (2009) Environmental change controls postglacial forest dynamics through interspecific differences in life-history traits. Ecology 90:2149–2160. doi:10.1890/08-1136.1

Lacourse T, Mathewes R, Fedje D (2005) Late-glacial vegetation dynamics of the Queen Charlotte Islands and adjacent continental shelf, British Columbia, Canada. Palaeogeogr Palaeoclimatol Palaeoecol 226:36–57. doi:10.1016/j.palaeo.2005.05.003

Leffler SR (1979) A new subspecies of *Cicindela bellissima* from Northwestern Washington (Coleoptera: Cicindelidae). Coleopt Bull 33:465–472. doi:10.2307/4000086

Legendre P, Galzin R, Harmelin-Vivien ML (1997) Relating behavior to habitat: solutions to the fourth-corner problem. Ecology 78:547–562. doi:10.1890/0012-9658(1997)078[0547:RBTHST]2.0.CO;2

Leopold EB, Boyd R (1999) An ecological history of old prairie areas, southwestern Washington. In: Boyd R (ed) Indians, fire and the land. Oregon State University Press, Corvallis, pp 139–166

Leopold EB, Nickman R, Hedges JI, Ertel RJ (1982) Pollen and lignin records of late-Quaternary vegetation, Lake Washington. Science 218:1305–2307. doi:10.1126/science.218.4579.1305

Lepofsky D, Lertzman K (2008) Documenting ancient plant management in the northwest of North America. Botany 86:129–145. doi:10.1139/B07-094

Lepofsky D, Lertzman KP, Hallett D, Mathewes R (2005) Climate change and culture change on the Southern Coast of British Columbia 2,400-1,200 cal. BP: an hypothesis. Am Antiquity 70:267–293

Lian OB, Mathewes RW, Hicock SR (2001) Palaeoenvironmental reconstruction of the Port Moody Interstade, a nonglacial interval in southwestern British Columbia at about 18 000 C-14 years BP. Can J Earth Sci 38:943–952. doi:10.1139/cjes-38-6-943

Lindsey AW (1939) A new species of Hesperia (Lepidoptera, Hesperiidae). Ann Entomol Soc Am 32:171–172

Littell JS, Peterson DL, Tjoelker M (2008) Douglas-fir growth in mountain ecosystems: water limits tree growth from stand to region. Ecol Monogr 78:349–368. doi:10.1890/07-0712.1

Liu Z, Otto-Bliesner BL, He F et al (2009) Transient simulation of last deglaciation with a new mechanism for Bølling-Allerød warming. Science 325:310–314. doi:10.1126/science.1171041

Logan RL, Schuster RL (1991) Lakes divided: The origin of Lake Crescent and Lake Sutherland, Clallam County, Washington. Washington Geol 19:38–42

Long WA (1975) Neoglaciation in the Northeastern Olympic Mountains, Washington. Unpublished Report

Long CJ, Whitlock C (2002) Fire and vegetation history from the coastal rain forest of the western Oregon Coast Range. Quat Res 58:215–225. doi:10.1006/qres.2002.2378

Long CJ, Whitlock C, Bartlein PJ (2007) Holocene vegetation and fire history of the Coast Range, western Oregon, USA. Holocene 17:917–926. doi:10.1177/0959683607082408

Loy WG, Allan S, Buckley AR, Meacham JE (2001) Atlas of Oregon, 2nd edn. Oregon State University Press, Corvallis

Lucas JD, Lacourse T (2013) Holocene vegetation history and fire regimes of *Pseudotsuga menziesii* forests in the Gulf Islands National Park Reserve, southwestern British Columbia, Canada. Quat Res 79:366–376. doi:10.1016/j.yqres.2013.03.001

Luken JO, Fonda RW (1983) Nitrogen accumulation in a chronosequence of red alder communities along the Hoh River, Olympic National Park, Washington. Can J Forest Res 13:1228–1237. doi:10.1139/x83-161

Lyle M, Wilkins D, Barron J, Heusser L (2003) North Pacific Sea surface temperature, Western U.S. Vegetation, and the demise of the Miocene Rocky Mountain Monsoon. AGU Fall Meeting Abstracts- 1:1185

Lyle M, Barron J, Bralower TJ et al (2008) Pacific Ocean and Cenozoic evolution of climate. Rev Geophys 46. doi:10.1029/2005RG000190

Lyman RL (1995) Determining when rare (zoo-)archaeological phenomena are truly absent. J Archaeol Method Theory 2:369–424. doi:10.1007/BF02229004

Lynott RE, Cramer OP (1966) Detailed analysis of 1962 Columbus Day Windstorm in Oregon and Washington. Mon Weather Rev 94:105–117. doi:10.1175/1520-0493(1966)094<0105:DAOTCD>2.3.CO;2

Lysak T, Ross DW, Maguire DA, Overhulser DL (2006) Predicting spruce weevil damage in Sitka spruce in the northern Oregon Coast Range. West J Appl Forest 21:159–164

Mantua NJ, Hare SR, Zhang Y, et al. (1997) A Pacific interdecadal climate oscillation with impacts on salmon production. Bull Am Meteorol Soc 78:1069–1079. doi:10.1175/1520-0477(1997)078<1069:APICOW>2.0.CO;2

Mapes L (2009) Breaking ground: the Lower Elwha Klallam tribe and the unearthing of Tse-whit-zen Village. University of Washington Press, Seattle

Marcott SA, Shakun JD, Clark PU, Mix AC (2013) A reconstruction of regional and global temperature for the past 11,300 Years. Science 339:1198–1201. doi:10.1126/science.1228026

Marcus WA, Meacham JE, Rodman AW, Steingisser AY (2012) Atlas of Yellowstone, 1st edn. University of California Press, Berkeley

Mass C (1982) The topographically forced diurnal circulations of western Washington State and their influence on precipitation. Mon Weather Rev 110:170–183. doi:10.1175/1520-0493(1982)110<0170:TTFDCO>2.0.CO;2

Mass C, Dotson B (2010) Major extratropical cyclones of the northwest United States: historical review, climatology, and synoptic environment. Mon Weather Rev 138:2499–2527. doi:10.1175/2010MWR3213.1

Mathewes RW (1973) A palynological study of post-glacial vegetation changes in the University Research Forest, southwestern British Columbia. Can J Bot 51:2085–2103. doi:10.1139/b73-271

Mathewes RW (1985) Paleobotanical evidence for climatic change in southern British Columbia during Late-Glacial and Holocene time. In: Harington CR (ed) Climatic change in Canada 5: critical periods in the Quaternary climatic history of Northern North America. Canada National Museum of Natural Sciences, Ottawa, pp 397–422 (Syllogeus 55)

Mathewes RW (1993) Evidence for Younger Dryas-age cooling on the North Pacific coast of America. Quat Sci Rev 12:321–331. doi:10.1016/0277-3791(93)90040-S

Mathewes RW, Heusser LE (1981) A 12,000 year palynological record of temperature and precipitation trends in southwestern British Columbia. Can J Bot 59:707–710

Mathewes RW, Rouse GE (1975) Palynology and paleoecology of postglacial sediments from the lower Fraser River Canyon of British Columbia. Can J Earth Sci 12:745–756. doi:10.1139/e75-065

Mathewes RW, Borden CE, Rouse GE (1972) New radiocarbon dates from the Yale area of the Lower Fraser River Canyon, British Columbia. Can J Earth Sci 9:1055–1057. doi:10.1139/e72-089

Mathewes RW, Huesser LE, Patterson RT (1993) Evidence for a younger dryas like cooling event on the British Columbia coast. Geology 21:101–104. doi:10.1130/0091-7613(1993)021<0101:EF AYDL>2.3.CO;2

McCloskey SPJ, Daniels LD, McLean JA (2009) Potential impacts of climate change on western hemlock looper outbreaks. Northwest Sci 83:225–238. doi:10.3955/046.083.0306

McLachlan J, Brubaker LB (1995) Local and regional vegetation change on the northeastern Olympic Peninsula during the Holocene. Can J Bot 73:1618–1627. doi:10.1139/b95-175

McMahon SM, Harrison SP, Armbruster WS et al (2011) Improving assessment and modelling of climate change impacts on global terrestrial biodiversity. Trends Ecol Evol 26:249–259. doi:10.1016/j.tree.2011.02.012

McMillan AD, Hutchinson I (2002) When the mountain dwarfs danced: Aboriginal traditions of paleoseismic events along the Cascadia Subduction Zone of western North America. Ethnohistory 49:41–68. doi:10.1215/00141801-49-1-41

McNulty T (2009) Olympic National Park: a natural history, 2nd edn. University of Washington Press, Seattle

McPhail JD (1967) Distribution of freshwater fishes in western Washington. Northwest Sci 41:1–11

Means JE (1990) Tsuga mertensiana (Bong.) Carr.: mountain hemlock. In: Burns RM, Honkala BH (eds) Silvics of North America: volume I, conifers. USDA Forest Service, Washington, pp 623–634

Menne MJ, Williams CN, Vose RS (2009) The United States historical climatology network monthly temperature data—version 2. Bull Am Meteorol Soc 90:993–1107

Menounos B, Clague JJ, Osborn G et al (2008) Western Canadian glaciers advance in concert with climate change circa 4.2 ka. Geophys Res Lett. doi:10.1029/2008GL033172

Menounos B, Osborn G, Clague JJ, Luckman BH (2009) Latest Pleistocene and Holocene glacier fluctuations in western Canada. Quat Sci Rev 28:2049–2074. doi:10.1016/j.quascirev.2008.10.018

Miller PA, Giesecke T, Hickler T et al (2008) Exploring climatic and biotic controls on Holocene vegetation change in Fennoscandia. J Ecol 96:247–259. doi:10.1111/j.1365-2745.2007.01342.x

Millington JDA, Perry GLW, Malamud BD (2006) Models, data and mechanisms: quantifying wildfire regimes, vol 261. Geological Society, London (Special Publications), pp 155–167. doi:10.1144/GSL.SP.2006.261.01.12

Minckley T, Whitlock C (2000) Spatial variation of modern pollen in Oregon and southern Washington, USA. Rev Palaeobot Palynol 112:97–123. doi:10.1016/S0034-6667(00)00037-3

Minder JR, Durran DR, Roe GH, Anders AM (2008) The climatology of small-scale orographic precipitation over the Olympic Mountains: patterns and processes. Q J Roy Meteorol Soc 134:817–839. doi:10.1002/qj.258

Minder JR, Roe GH, Montgomery DR (2009) Spatial patterns of rainfall and shallow landslide susceptibility. Water Resour Res. doi:10.1029/2008WR007027

Moeur M, Ohmann JL, Kennedy RE et al (2011) Northwest Forest Plan-the first 15 years (1994–2008): status and trends of late-successional and old-growth forests. USDA Forest Service General Technical Report PNW-GTR-853, Portland

Monsen KJ, Blouin MS (2003) Genetic structure in a montane ranid frog: restricted gene flow and nuclear-mitochondrial discordance. Mol Ecol 12:3275–3286

Montgomery DR (2002) Valley formation by fluvial and glacial erosion. Geology 30:1047–1050. doi:10.1130/0091-7613(2002)030<1047:VFBFAG>2.0.CO;2

Montgomery DR, Abbe TB (2006) Influence of logjam-formed hard points on the formation of valley-bottom landforms in an old-growth forest valley, Queets River, Washington, USA. Quat Res 65:147–155. doi:10.1016/j.yqres.2005.10.003

Moritz MA, Morais ME, Summerell LA et al (2005) Wildfires, complexity, and highly optimized tolerance. Proc Natl Acad Sci U S A 102:17912–17917. doi:10.1073/pnas.0508985102

Moss M (2011) Northwest Coast: archaeology as deep history. Society for American Archaeology, Washington

Mustoe GE, Harington CR, Morlan RE (2005) Cedar hollow, an early Holocene faunal site from Whidbey Island, Washington. West North Am Nat 65:429–440

Nagorsen DW, Keddie G (2000) Late Pleistocene mountain goats (Oreamnos americanus) from Vancouver Island: biogeographic implications. J Mamm 81:666–675

Nakawatase JM, Peterson DL (2006) Spatial variability in forest growth—climate relationships in the Olympic Mountains, Washington. Can J Forest Res 36:77–91. doi:10.1139/X05-224

Nelson AR, Kelsey HM, Witter RC (2006) Great earthquakes of variable magnitude at the Cascadia subduction zone. Quat Res 65:354–365. doi:10.1016/j.yqres.2006.02.009

Nielson M, Lohman K, Sullivan J (2001) Phylogeography of the Tailed Frog (*Ascaphus truei*): Implications for the biogeography of the Pacific Northwest. Evolution 55:147–160. doi:10.2307/2640697

Niinemets U, Valladares F (2006) Tolerance to shade, drought, and waterlogging of temperate Northern Hemisphere trees and shrubs. Ecol Monogr 76:521–547. doi:10.1890/0012-9615(2006)076[0521:TTSDAW]2.0.CO;2

Nolin AW, Daly C (2006) Mapping "at risk" snow in the Pacific Northwest. J Hydrometeor 7:1164–1171. doi:10.1175/JHM543.1

North Greenland Ice Core Project members (2004) High-resolution record of Northern Hemisphere climate extending into the last interglacial period. Nature 431:147–151

O'Connell LM, Ritland K, Thompson SL (2008) Patterns of post-glacial colonization by western redcedar (*Thuja plicata*, Cupressaceae) as revealed by microsatellite markers. Botany 86:194–203. doi:10.1139/B07-124

O'Connor JE, Jones MA, Haluska TL (2003) Flood plain and channel dynamics of the Quinault and Queets Rivers, Washington, USA. Geomorphology 51:31–59. doi:10.1016/S0169-555X(02)00324-0

Osborn G, Menounos B, Ryane C, et al. (2012) Latest Pleistocene and Holocene glacier fluctuations on Mount Baker, Washington. Quat Sci Rev 49:33–51. doi:10.1016/j.quascirev.2012.06.004

Palmer S, Walker I, Heinrichs M, et al. (2002) Postglacial midge community change and Holocene palaeotemperature reconstructions near treeline, southern British Columbia (Canada). J Paleolimnol 28:469–490. doi:10.1023/A:1021644122727

Pazzaglia FJ, Brandon MT (2001) A fluvial record of long-term steady-state uplift and erosion across the Cascadia forearc high, western Washington State. Am J Sci 301:385–431

Pellatt MG, Hebda RJ, Mathewes RW (2001) High-resolution Holocene vegetation history and climate from Hole 1034B, ODP leg 169S, Saanich Inlet, Canada. Mar Geol 174:211–226

Pellatt MG, Mathewes RW, Clague JJ (2002) Implications of a late-glacial pollen record for the glacial and climatic history of the Fraser Lowland, British Columbia. Palaeogeogr Palaeoclimatol Palaeoecol 180:147–157. doi:10.1016/S0031-0182(01)00426-6

Pellatt MG, Mathewes RW, Clague JJ (2004) Reply to comments on "Implications of a late-glacial pollen record for the glacial and climatic history of the Fraser Lowland, British Columbia." Palaeogeogr Palaeoclimatol Palaeoecol 203:343–346. doi:10.1016/S0031-0182(03)00723-5

Petersen KL, Mehringer Jr. PJ, Gustafson CE (1983) Late-glacial vegetation and climate at the Manis Mastodon site, Olympic Peninsula, Washington. Quat Res 20:215–231. doi:10.1016/0033-5894(83)90078-9

Peterson DL (1998) Climate, limiting factors and environmental change in high-altitude forests of Western North America. In: Beniston M, Innes JL (eds) The impacts of climate variability on forests. Springer-Verlag, Berlin, pp 191–208

Peterson DL, Schreiner EG, Buckingham NM (1997a) Gradients, vegetation and climate: spatial and temporal dynamics in the Olympic Mountains, USA. Glob Ecol Biogeogr Lett 6:7–17

Peterson EB, Peterson NM, Weetman GF, Martin PJ (1997b) Ecology and management of Sitka Spruce: emphasizing its natural range in British Columbia. University of British Columbia Press, Vancouver

Pickford SG, Fahnestock G, Ottmar R (1980) Weather, fuel, and lightning fires in Olympic National Park. Northwest Sci 54:92–105

Pinchot G (1947) Breaking New Ground. Harcourt, Brace, and Co, New York

Pisaric MFJ (2002) Long-distance transport of terrestrial plant material by convection resulting from forest fires. J Paleolimnol 28:349–354. doi:10.1023/A:1021630017078

Porter SC, Swanson TW (1998) Radiocarbon age constraints on rates of advance and retreat of the Puget lobe of the Cordilleran Ice Sheet during the last glaciation. Quat Res 50:205–213

Prichard SJ, Gedalof Z, Oswald WW, Peterson DL (2009) Holocene fire and vegetation dynamics in a montane forest, North Cascade Range, Washington, USA. Quat Res 72:57–67. doi:10.1016/j.yqres.2009.03.008

Reagan AB (1909) Some notes on the Olympic Peninsula, Washington. Trans Kans Acad Sci 22:131–238

Rehn JAG (1952) Two new Melanoploid genera (Orthoptera: Acrididae: Cyrtacanthacridinae) from the western United States. Trans Am Entomol Soc (1890-) 78:101–115. doi:10.2307/25077646

Reimer PJ, Baillie MGL, Bard E, et al (2009) INTCAL09 and MARINE09 radiocarbon age calibration curves, 0-50,000 years cal BP. Radiocarbon 51:1111–1150

Richart CH, Hedin M (2013) Three new species in the harvestmen genus Acuclavella (Opiliones, Dyspnoi, Ischyropsalidoidea), including description of male Acuclavella quattuor Shear, 1986. Zookeys 19–68. doi:10.3897/zookeys.311.2920

Ricotta C, Arianoutsou M, Díaz-Delgado R, et al. (2001) Self-organized criticality of wildfires ecologically revisited. Ecol Model 141:307–311. doi:10.1016/S0304-3800(01)00272-1

Riedel JL, Clague JJ, Ward BC (2010) Timing and extent of early marine oxygen isotope stage 2 alpine glaciation in Skagit Valley, Washington. Quat Res 73:313–323. doi:10.1016/j.yqres.2009.10.004

Rochefort RM, Little RL, Woodward A, Peterson DL (1994) Changes in sub-alpine tree distribution in western North America: a review of climatic and other causal factors. Holocene 4:89–100. doi:10.1177/095968369400400112

Rosenberg SA, Walker IR, Mathewes RW, Hallett DJ (2004) Midge-inferred Holocene climate history of two subalpine lakes in southern British Columbia, Canada. Holocene 14:258–271

Schalk RF (1988) The evolution and diversification of native land use systems on the Olympic Peninsula: a research design. Contract No. CX-900-4-E075. National Park Service, Port Angeles

Schmidt RL (1960) Factors controlling the distribution of Douglas-fir in coastal British Columbia. Q J Forest 54:155–160

Schoonmaker PK, Foster DR (1991) Some implications of paleoecology for contemporary ecology. Bot Rev 57:204–245. doi:10.1007/BF02858563

Schreiner EG (1994) Subalpine and alpine plant communities. In: Houston DB, Schreiner EG, Moorhead BB (eds) Mountain Goats in Olympic National Park: Biology and management of an Introduced Species. Olympic National Park, National Park Service, Port Angeles, Washington, pp 242–250

Schreiner EG, Burger JE (1994) Photographic comparisons: a qualitative appraisal of the influence of climate and disturbances on vegetation, 1915–1990. In: Houston DB, Schreiner EG, Moorhead BB (eds) Mountain goats in Olympic National Park: biology and management of an introduced species. National Park Service, Port Angeles, Washington.

Schreiner EG, Gracz MB, Kaye TN et al (1994) Rare plants. In: Houston DB, Schreiner EG, Moorhead BB (eds) Mountain goats in Olympic National Park: biology and management of an introduced species. National Park Service, Port Angeles, Washington.

Schreiner E, Krueger K, Happe P, Houston D (1996) Understory patch dynamics and ungulate herbivory in old-growth forests of Olympic National Park, Washington. Can J Forest Res 26:255–265. doi:10.1139/x26-029

Schultz S (1994) Review of the historical evidence relating to mountain goats in the Olympic Mountains before 1925. In: Houston DB, Schreiner EG, Moorhead BB (eds) Mountain goats in Olympic National Park: biology and management of an introduced species. Olympic National Park, National Park Service, Port Angeles, Washington, pp 242–250

Schuster RL, Logan RL, Pringle PT (1992) Prehistoric rock avalanches in the Olympic Mountains, Washington. Science 258:1620–1621. doi:10.2307/2882062

Serder CF (1999) Description, analysis and impacts of the Grouse Creek landslide, Jefferson County, Washington, 1997–98. M.S. Thesis, The Evergreen State College

Shafer ABA, Cullingham CI, Cote SD, Coltman DW (2010) Of glaciers and refugia: a decade of study sheds new light on the phylogeography of northwestern North America. Mol Ecol 19:4589–4621. doi:10.1111/j.1365-294X.2010.04828.x

Shear WA, Leonard WP (2003) Microlympiidae, a new milliped family from North America, and Microlympia echina, new genus and species (Diplopoda: Chordeumatida: Brannerioidea). Zootaxa 243:1–11

Shelley RM, Shear WA (2006) Leonardesmus injucundus, n. gen., n. sp., an aromatic, small-bodied milliped from Washington State, U. S. A., and a revised account of the family Nearctodesmidae (Polydesmida). Zootaxa 1176:1–16

Simard M, Pinto N, Fisher JB, Baccini A (2011) Mapping forest canopy height globally with spaceborne lidar. J Geophys Res: Biogeosci 116:n/a-n/a. doi:10.1029/2011JG001708

Simpson GG (1964) Species density of North American recent mammals. Syst Zool 13:57–73. doi:10.2307/2411825

Soll JA (1994) Seed number, germination and first year survival of subalpine fir (abies lasiocarpa) in subalpine meadows of the Northeastern Olympic mountains. M.S. Thesis, University of Washington

Spear RW (1989) Late-Quaternary history of high-elevation vegetation in the White Mountains of New Hampshire. Ecol Monogr 59:125–151. doi:10.2307/2937283

Spicer RC (1989) Recent variations of Blue Glacier, Olympic Mountains, Washington, USA. Arctic Alpine Res 21:1–21

Steele CA, Storfer A (2007) Phylogeographic incongruence of codistributed amphibian species based on small differences in geographic distribution. Mol Phylogenet Evol 43:468–479. doi: 10.1016/j.ympev.2006.10.010

Stein JK (2000) Exploring Coast Salish Prehistory. University of Washington Press, Seattle

Stolnack S, Naiman RJ (2010) Patterns of conifer establishment and vigor on montane river floodplains in Olympic National Park, Washington, USA. Can J Forest Res 40:410–422. doi:10.1139/X09-200

Sugimura WY, Sprugel DG, Brubaker LB, Higuera PE (2008) Millennial-scale changes in local vegetation and fire regimes on Mount Constitution, Orcas Island, Washington, USA, using small hollow sediments. Can J Forest Res 38:539–552. doi:10.1139/X07-186

Sugita S (1990) Palynological records of forest disturbance and development in the Mountain Meadows watershed, Mt. Rainier, Washington. Ph. D. Thesis, University of Washington

Sugita S, Tsukada M (1982) The vegetation history in western North America 1. Mineral and Hall lakes. Japanese J Ecol 32:499–516

Sutton DA (1992) Tamias amoenus. Mamm Species 390:1–8. doi:10.2307/3504206

Swanson FJ, Kratz TK, Caine N, Woodmansee RG (1988) Landform effects on ecosystem patterns and processes. BioScience 38:92–98. doi:10.2307/1310614

Swanson ME, Franklin JF, Beschta RL, et al. (2010) The forgotten stage of forest succession: early-successional ecosystems on forest sites. Front Ecol Environ 9:117–125. doi:10.1890/090157

Tabor RW (1975) Guide to the geology of Olympic National Park. University of Washington Press, Seattle

Tattersall AM (1999) Changes in the distribution of selected conifer taxa in the Pacific Northwest during the last 20,000 years. M.S. Thesis, University of Oregon

Taylor AH (1990) Disturbance and persistence of Sitka spruce (Picea sitchensis (Bong) Carr) in coastal forests of the Pacific Northwest, North America. J Biogeogr 17:47–58. doi:10.2307/2845187

Thackray GD (2001) Extensive early and middle Wisconsin glaciation on the western Olympic Peninsula, Washington, and the variability of Pacific moisture delivery to the northwestern United States. Quat Res 55:257–270. 10.1006/qres.2001.2220

Thackray GD (2008) Varied climatic and topographic influences on Late Pleistocene mountain glaciation in the western United States. J Quat Sci 23:671–681. doi:10.1002/jqs.1210

Thompson RS, Anderson KH, Bartlein PJ (1999) Atlas of relations between climatic parameters and distributions of important trees and shrubs in North America. United States Geological Survey Professional Paper 1650 A & B

Thompson RS, Shafer SL, Strickland LE, et al. (2003) Quaternary vegetation and climate change in the western United States: Developments, perspectives, and prospects. In: Gillespie AR, Porter SC, Atwater BF (eds) The Quaternary period in the United States. Elsevier, pp 403–426

Tinner W, Conedera M, Ammann B, Lotter AF (2005) Fire ecology north and south of the Alps since the last ice age. Holocene 15:1214–1226. doi:10.1191/0959683605hl892rp

Tisch EL (2001) *Corallorhiza maculata* var. *ozettensis* (Orchidaceae), a new coral-root from coastal Washington. Madrono 48:40–42

Tsukada M, Sugita S (1982) Late Quaternary dynamics of pollen influx at Mineral Lake, Washington. Bot Mag (Tokyo) 95:401–418

Tsukada M, Sugita S, Hibbert DM (1981) Paleoecology in the Pacific Northwest I. Late Quaternary vegetation and climate. Int Ver Theor Angew 21:730–737

Vacco D, Clark P, Mix A et al (2005) A speleothem record of Younger Dryas cooling, Klamath Mountains, Oregon, USA. Quat Res 64:249–256. doi:10.1016/j.yqres.2005.06.008

Van der Hammen T (1974) The Pleistocene changes of vegetation and climate in tropical South America. J Biogeogr 1:3–26. doi:10.2307/3038066

Van Pelt R, O'Keefe TC, Latterell JJ, Naiman RJ (2008) Riparian forest stand development along the Queets River in Olympic National Park, Washington. Ecol Monogr 76:277–298. doi:10.1890/05-0753

Verts BJ, Carraway LN (2000) Thomomys mazama. Mamm Species 641:1–7. doi:10.1644/1545-1410(2000)641<0001:TM>2.0.CO;2

Wainman N, Mathewes RW (1987) Forest history of the last 12000 years based on plant macrofossil analysis of sediment from Marion Lake, southwestern British Columbia. Can J Bot 65:2179–2187

Wang T, Hamann A, Spittlehouse D, Murdock TN (2012) ClimateWNA—high-resolution spatial climate data for western North America. J Appl Meteorol Climatol 61:16–29. doi:10.1175/JAMC-D-11-043.1

Ward K, Shoal R, Aubry C (2006) Whitebark Pine in Washington and Oregon: A Synthesis of current studies and historical data. USDA Forest Service Region 6, Portland

Washington Division of Geology and Earth Resources staff (2008) Digital geology of Washington state at 1:100,000 scale

Waters MR, Stafford TW (2007) Redefining the age of Clovis: implications for the peopling of the Americas. Science 315:1122–1126. doi:10.1126/science.1137166

Waters MR, Stafford TW, McDonald HG et al (2011) Pre-Clovis mastodon hunting 13,800 years ago at the Manis Site, Washington. Science 334:351–353. doi:10.1126/science.1207663

Webb TI (1986) Is vegetation in equilibrium with climate? How to interpret late-Quaternary pollen data. Vegetatio 67:75–91. doi:10.2307/20146308

Wegmann KW, Pazzaglia FJ (2002) Holocene strath terraces, climate change, and active tectonics: the Clearwater River basin, Olympic Peninsula, Washington State. Geol Soc of Am Bull 114:731–744. doi:10.1130/0016-7606(2002)114<0731:HSTCCA>2.0.CO;2

Weisberg PJ (2004) Importance of non-stand-replacing fire for development of forest structure in the Pacific Northwest, USA. Forest Sci 50:245–258

Weisberg PJ, Swanson FJ (2003) Regional synchroneity in fire regimes of western Oregon and Washington, USA. Forest Ecol Manage 172:17–28

Weiser A, Lepofsky D (2009) Ancient land use and management of Ebey's Prairie, Whidbey Island, Washington. J Ethnobiol 29:184–212

Welch CK (2008) Historical biogeography of Pacific Northwest (USA) fossorial mammals and Australian treecreepers. Ph. D. Thesis, University of Washington

Wendel R, Zabowski D (2010) Fire history within the lower Elwha River watershed, Olympic National Park, Washington. Northwest Sci 84:88–97

Wetzel S, Fonda R (2000) Fire history of Douglas-fir forests in the Morse Creek drainage of Olympic National Park, Washington. Northwest Sci 74:263–279

Whitlock C (1992) Vegetational and climatic history of the Pacific Northwest during the last 20,000 years: implications for understanding present day biodiversity. Northwest Environ J 8:5–28

Whitlock C, Marlon J, Briles C et al (2008) Long-term relations among fire, fuel, and climate in the northwestern US based on lake-sediment studies. Int J Wildland Fire 17:72–83. doi:10.1071/WF07025

Wiggins DA (1999) The peninsula effect on species diversity: a reassessment of the avifauna of Baja California. Ecography 22:542–547

Willis KJ, Bailey RM, Bhagwat SA, Birks HJB (2010) Biodiversity baselines, thresholds and resilience: testing predictions and assumptions using palaeoecological data. Trends Ecol Evol 25:583–591. doi:10.1016/j.tree.2010.07.006

Willmott CJ, Rowe CM, Mintz Y (1985) Climatology of the terrestrial seasonal water cycle. J Climatol 5:589–606. doi: 10.1002/joc.3370050602

Wilson JR, Bartholomew MJ, Carson RJ (1979) Late Quaternary faults and their relationship to tectonism in the Olympic Peninsula, Washington. Geology 7:235–239. doi:10.1130/0091-7613(1979)7<235:LQFATR>2.0.CO;2

Wilson MC, Kenady SM, Schalk RF (2009) Late Pleistocene Bison antiquus from Orcas Island, Washington, and the biogeographic importance of an early postglacial land mammal dispersal corridor from the mainland to Vancouver Island. Quat Res 71:49–61. doi:10.1016/j.yqres.2008.09.001

Winchester NN, Ring RA (1999) The biodiversity of arthropods from northern temperate ancient coastal rainforests: conservation lessons from the high canopy. Selbyana 20:268–275

Winter LE, Brubaker LB, Franklin JF, et al. (2002) Initiation of an old-growth Douglas-fir stand in the Pacific Northwest: a reconstruction from tree-ring records. Can J Forest Res 32:1039–1056. doi:10.1139/X02-031

Witter RC, Givler RW, Carson RJ (2008) Two post-glacial earthquakes on the Saddle Mountain West Fault, Southeastern Olympic Peninsula, Washington. Bull Seismol Soc Am 98:2894–2917. doi:10.1785/0120080127

Wolfe JA (1985) Distribution of major vegetational types during the Tertiary. In: Sundquist ET, Broecker WS (eds) Geophysical monograph series. American Geophysical Union, Washington, DC, pp 357–375

Woodward A, Schreiner EG, Houston DB, Moorhead BB (1994) Ungulate-forest relationships in Olympic National Park—Retrospective exclosure studies. Northwest Sci 68:97–110

Woodward A, Schreiner EG, Silsbee DG (1995) Climate, geography and tree establishment in sub-alpine meadows of the Olympic Mountains, Washington, USA. Arctic Alpine Res 27:217–225

Wright HE, Mann DH, Glaser PH (1984) Piston corers for peat and lake sediments. Ecology 65:657. doi:10.2307/1941430

Yake B (2005) Butterflies of the southern Olympics: The results of surveys from 2002–2004 with special focus on 2004. A report prepared for the Olympic National Park

Yeats RS (2004) Living with Earthquakes in the Pacific Northwest, 2nd edn. Oregon State University Press, Corvallis

Zachos J, Pagani M, Sloan L et al (2001) Trends, rhythms, and aberrations in global climate 65 Ma to present. Science 292:686–693. doi:10.1126/science.1059412

Zdanowicz CM, Zielinski GA, Germani MS (1999) Mount Mazama eruption: Calendrical age verified and atmospheric impact assessed. Geology 27:621–624. doi:10.1130/0091-7613

Ziegltrum GJ (2004) Efficacy of black bear supplemental feeding to reduce conifer damage in western Washington. J Wildlife Manage 68:470-474. doi:10.2193/0022-541X(2004)068[0470:EOBBSF]2.0.CO;2

Zolbrod AN, Peterson DL (1999) Response of high-elevation forests in the Olympic Mountains to climatic change. Can J Forest Res 29:1966–1978

Index

© Springer International Publishing Switzerland 2015
D. G. Gavin, L. B. Brubaker, *Late Pleistocene and Holocene Environmental Change on the Olympic Peninsula, Washington,* Ecological Studies 222,
DOI 10.1007/978-3-319-11014-1